Nelson Maths

2

W0036218

Pupil Book

Karen Morrison
Lisa Greenstein

OXFORD
UNIVERSITY PRESS

OXFORD
UNIVERSITY PRESS

Great Clarendon Street, Oxford, OX2 6DP, United Kingdom

Oxford University Press is a department of the University of Oxford.

It furthers the University's objective of excellence in research, scholarship, and education by publishing worldwide. Oxford is a registered trade mark of Oxford University Press in the UK and in certain other countries.

First published 2022

British Library Cataloguing in Publication Data

Data available

ISBN: 978-1-382-01000-9

1 3 5 7 9 10 8 6 4 2

Paper used in the production of this book is a natural, recyclable product made from wood grown in sustainable forests. The manufacturing process conforms to the environmental regulations of the country of origin.

Printed in Great Britain by Bell and Bain Ltd, Glasgow

Acknowledgements

The publisher and authors would like to thank the following for permission to use photographs and other copyright material:

Cover: Matthieu Nivesse. Photos: **p26(l):** Yingna Cai/Shutterstock; **p26(ml):** nrg123/Shutterstock; **p26(r):** Marlene Ford / Alamy Stock Photo; **p27(br):** Africa Studio/Shutterstock; **p27(bl):** pernsanitfoto/Shutterstock; **p27(tr):** MawardiBahar/Shutterstock; **p27(tl):** IS MODE/Shutterstock; **p55:** PhilipYb Studio/Shutterstock; **p111:** Ragnarock/Shutterstock.

Artwork by Liliana Perez, Q2A Media, Pantek Media, and OKS Prepress.

Every effort has been made to contact copyright holders of material reproduced in this book. Any omissions will be rectified in subsequent printings if notice is given to the publisher.

MIX
Paper from
responsible sources
FSC® C007785

Contents

Think maths

THINK MATHS prepares you to think mathematically, value mistakes, and learn maths with a growth mindset.

How I feel about maths

 **Think and share**

Read what these pupils said:

Everyone makes mistakes sometimes.

If I were good at maths, I wouldn't make mistakes.

Mistakes make me angry.

Making mistakes helps me to get better at maths.

How do you think each pupil feels about making a mistake?

1 When you do maths, you use your brain. There is no such thing as a maths brain. Everyone's brain looks a bit like this:

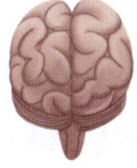

A human brain

a Which shape is most like the shape of the brain?

circle

square

triangle

b What kind of lines does it have?

Mistakes and your brain

When you use your brain, it makes connections and 'lights up'. This makes it grow.

Your brain grows most when you make mistakes! This means that mistakes help us to learn new things.

Which brain is making a mistake? How do you know?

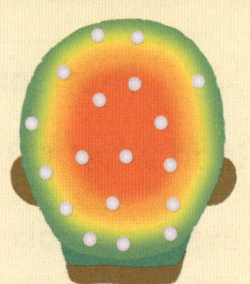

1 **a** How do you feel when you make a mistake?

When I make a mistake, I feel …

worried   excited   curious   sad

happy   interested   embarrassed

b Why?

2 Tell a partner why it is important to make mistakes.

3 Spot the mistake in each sentence. Find two different ways you could correct it. The first one is done for you.

3 less than 10 is 6.

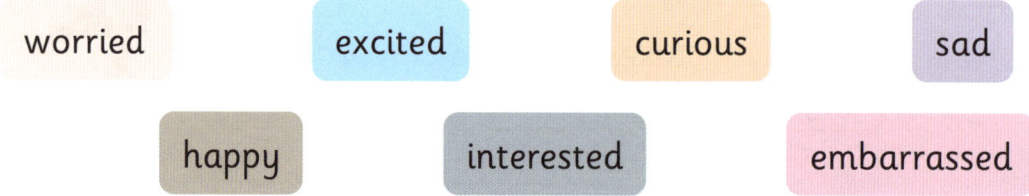

3 less than 10 is 7.
and
4 less than 10 is 6.

a A square has 3 sides.

b 2 more than 5 is 10.

c 5 o'clock is 2 hours before 3 o'clock.

➡ *Workbook page 4*

Different ways of working

There are many different ways to learn maths.

1 Look at the pictures.
What are the children doing?

2 a Match these statements to the pictures above.

We can work on maths together or alone.

Everyone can do maths.

Asking questions helps us to learn.

Working together helps us to solve problems.

b What else helps us to learn maths?

UNIT 2 Working with numbers

Count and show numbers

> 💭 **Think and share**
>
> Sometimes we do not need to count.
>
>
>
> How many handles?
>
>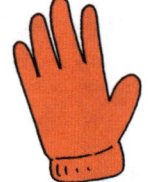
>
> How many fingers?

1 Guess how many dots in each ten frame.
Then count.

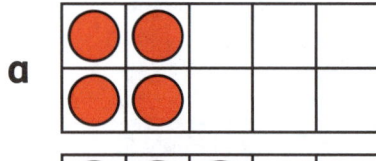

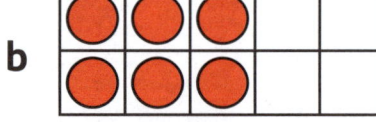

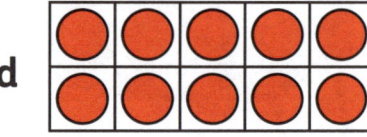

2 These ten frames all show the same number in different ways.

a How are the dots arranged in each ten frame?

b Which arrangement of dots is easiest for you to count?

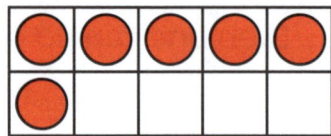

3 Choose a number between 1 and 10.
Show it in three different ways in a ten frame.

➡ *Workbook page 5*

Ordinal numbers

Ordinal numbers tell us the position of a person or object.

1st first	2nd second	3rd third	4th fourth	5th fifth	6th sixth	7th seventh	8th eighth	9th ninth	10th tenth

1   a   Who is first in line?

    b   Who is last?

    c   Which position comes before 3rd?

    d   Which position comes after 4th?

2   Which position are these pupils in?

Reshma   Priya   Zayed

Ali   Anisha   Adam

Count to 100

The numbers greater than 19 follow a rule.
You say a number for the **tens**, then a number for the **ones**, like this:

1	2	3	4	5	6	7	8	9	10
11	12	13	14	15	16	17	18	19	20
21	22	23	24	25	26	27	28	29	30
31	32	33	34	35	36	37	38	39	40
41	42	43	44	45	46	47	48	49	50
51	52	53	54	55	56	57	58	59	60
61	62	63	64	65	66	67	68	69	70
71	72	73	74	75	76	77	78	79	80
81	82	83	84	85	86	87	88	89	90
91	92	93	94	95	96	97	98	99	100

21
twenty-one

22
twenty-two

23
twenty-three

If there are no ones, you just say the tens:

30
thirty

40
forty

50
fifty

The numbers from 11 to 19 don't follow this rule.
You have to learn their names.

11
eleven

12
twelve

13
thirteen

14
fourteen

15
fifteen

16
sixteen

17
seventeen

18
eighteen

19
nineteen

1 Count all the way from 1 to 100.

2 Work with a partner.
Pick a number on the chart and say it.
Count the next 5 numbers, pointing to them on the chart.

3 Pick any number between 1 and 10.
Count on in tens.
How far can you go?

➡ *Workbook page 6*

Number names

We can read and write numbers like this: 21, 22, 23.
We can use words to write number names like this:
twenty-one
twenty-two
twenty-three

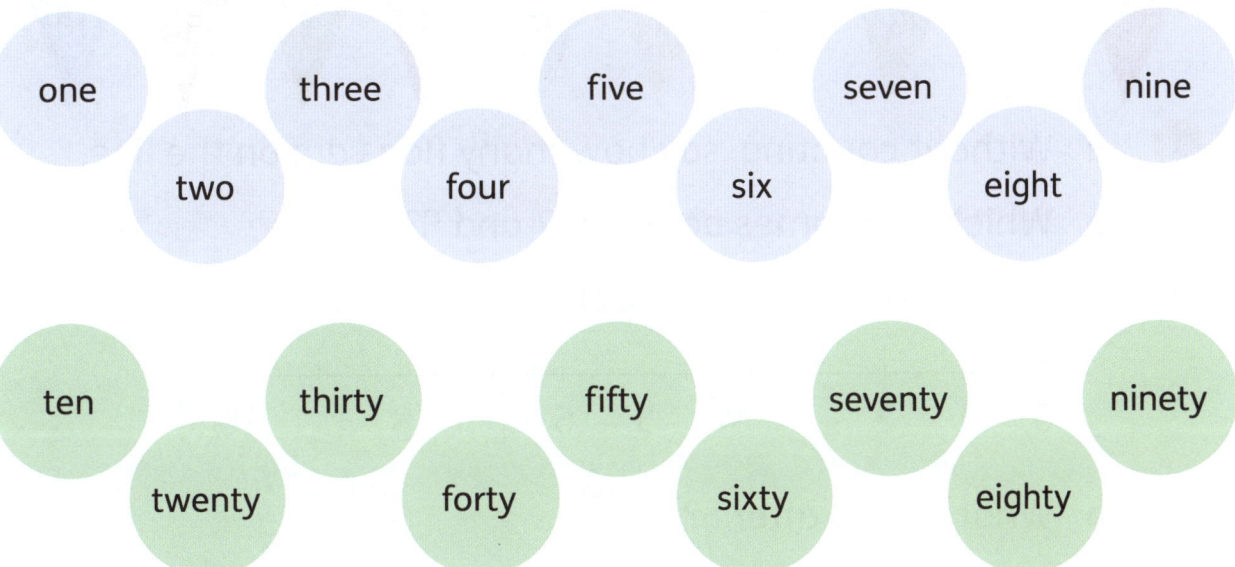

one
two
three
four
five
six
seven
eight
nine

ten
twenty
thirty
forty
fifty
sixty
seventy
eighty
ninety

1 Use the words to help you write the number names for these numbers.

a 41   **b** 75   **c** 54   **d** 60   **e** 89

2 Work with a partner.
Pick a number from the 100 chart.
Then point to the words to make the number name.

> Remember to put a hyphen (a short line) between the tens name and the ones name. So, we write 'twenty-one', not 'twenty one'.

➡ *Workbook page 7*

Use a number line

This is a **number line** from 0 to 20.

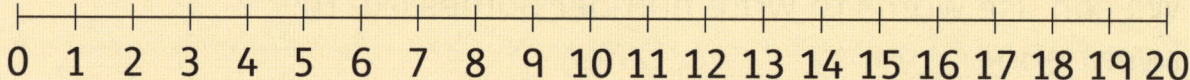

0 1 2 3 4 5 6 7 8 9 10 11 12 13 14 15 16 17 18 19 20

A number line can help us to count and **order** numbers.

1 2 3 4 5 6 7 8 9 10

1 a Without counting, say how many flags are on the line.

   b Which flag comes between 7 and 9?

2 a Which numbers are missing?

0 1 2 ☐ 4 5 6 ☐ 8 ☐ 10

   b What comes after 5?

   c What comes before 10?

3 This number line from 7 to 16 has two mistakes.
   What are they?

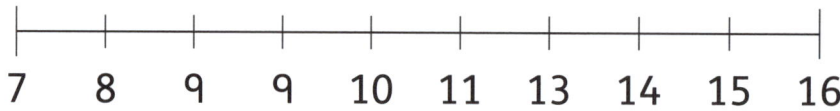

7 8 9 9 10 11 13 14 15 16

💡 **Problem solving**

First draw a line with 10 number places.

4 A number line shows 10 numbers.
   The middle numbers are 15 and 16.
   Draw the number line.

➡️ *Workbook page 8 and page 9*

Estimate and count

An **estimate** is a sensible guess about a number or a maths problem.
You can estimate an answer before you count or work it out.

1 a Estimate how many jelly beans there are.
Are there more than 10 or fewer than 10? More than 20? About 100?

b Explain how you estimated.

2 a Which bunch has the most flowers?
Estimate without counting.

b Count the red flowers.
Do you think there are more or fewer yellow flowers?
Count the yellow flowers to check.

c Estimate the number of purple flowers.
Then count.
How close was your estimate to the answer?

➡ *Workbook page 10*

Place value

Tens and ones

> ### 💭 Think and share
>
> We use the **digits** 0, 1, 2, 3, 4, 5, 6, 7, 8 and 9 to write all of our numbers.
>
> We know the **value** of a digit from its place in a number. This is called **place value**.
>
> Look at this **place-value table**.
>
> | Tens | Ones |
> |:---:|:---:|
> | ⫴ | 🟨 🟨 🟨 🟨 🟨 |
> | 3 | 5 |
>
> How many tens does it show?
> How many ones does it show?
> What number does it show?

1 Put blocks together to make towers of ten.
Make these numbers.

   a 20   **b** 30   **c** 40   **d** 50   **e** 100

2 **a** What do you notice about the number you made and the number of towers?

   b Why do numbers like 20, 30 and 40 have a zero after the first number? What does the zero tell us?

3 Say how many tens and how many ones.
Then say the number.

a

Tens	Ones
⫴	🟨 🟨 🟨 🟨

b

Tens	Ones
⫼	🟨 🟨 🟨 🟨 🟨 🟨

➡️ *Workbook page 11*

Make and break numbers

You can put small numbers together to make greater numbers.

Put 3 and 7 together.
Make 10.

You can break greater numbers into smaller numbers.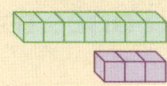

Break 10 into parts.
The two parts are 3 and 7.

1 How many different ways can you break 10 into two smaller numbers?
Use blocks to help you.

2 **a** Use blocks to show 19.
How many more blocks do you need to make 20?

b Make some other teen numbers using blocks.
How many more blocks do you need to make 20 each time?

We can write numbers in a place-value table.

Tens	Ones
2	3

3 Draw place-value tables to show these numbers.

a 1 ten and 9 ones   **b** 2 tens and 4 ones

c 4 tens and 8 ones   **d** 9 tens and 3 ones

💡 **Problem solving**

4 A shopkeeper sells packets of 10 pens.
She also sells one pen at a time.
One morning:
- 2 customers each buy 10 pens
- 12 customers each buy 1 pen.

How many does she sell altogether?

> Write the number of pens each person buys in a place-value table.

➡ *Workbook page 12*

Compose and decompose

Compose means put together.
Here are 3 tens and 5 ones.

You can put the tens and ones together.
They make 35.

Decompose means break apart.
Here is the number 27.

You can break it into tens and ones.
It has 2 tens and 7 ones.

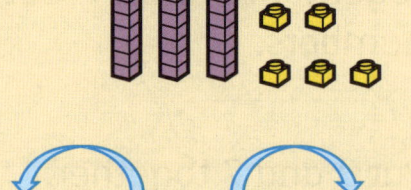

2 tens | 2 | 7 | 7 ones

1 How many tens, and how many ones?
Say and write each number in words.

a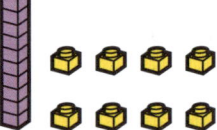

b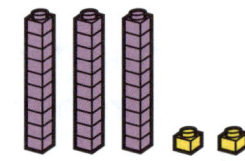

c

d

2 Use base-ten blocks to make these numbers.

a 29   b 48   c 76   d 94

3 Choose the correct value for the underlined red digit
in each number.

a <u>1</u>7   b <u>2</u>5   c <u>7</u>5   d <u>5</u>9
1 or 10?   2 or 20?   70 or 7?   5 or 50?

Problem solving

4 I am a number greater than 10 and less than 30.
My tens and ones places both have the same digit.
Which numbers could I be?

➡ *Workbook page 13*

Use the = sign

= means equal to.
When things are equal, they have the same value.

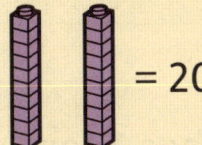 = 20

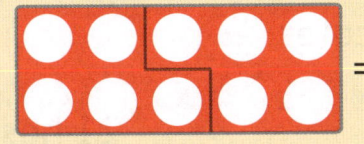 =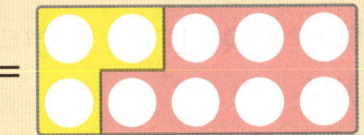

2 tens = 20   5 + 5 = 10   3 + 7 = 10

1 Look at each picture.
Explain why it shows equal amounts.

a
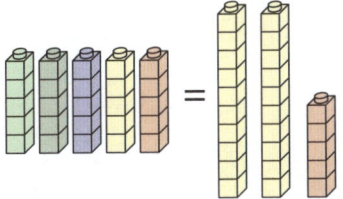

b
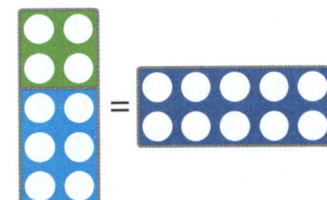

c
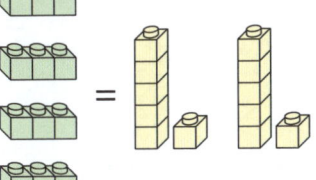

d

e
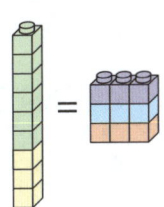

f
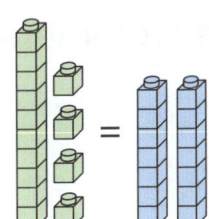

2 What is it equal to? The first one has been done for you.

a 12 = ___1 ten and 2 ones___

b 1 ten and 1 one = ☐

c 19 = _____

d 2 tens and 5 ones = ☐

e _____ = 30

f 5 tens and 0 ones = ☐

➡ *Workbook page 14*

Compare numbers

We use these signs to compare numbers:
The **< sign** means less than.
The **> sign** means greater than.

Look at the dots on these < and > signs:

18 < 24   24 > 18

The two dots always go next to the greater number.

You can also see the < and > signs on a number line:

0   18   24

When you compare numbers, first look at the digit with the greatest place value.

21 ☐ 15

21 has 2 tens   15 has 1 ten
The number with more tens is greater. 21 > 15

If the tens are the same, then compare the ones. 21 < 25

---

1   Choose the greater number.

a   7 or 4   b   9 or 14   c   11 or 20

2   Read the statements aloud.

a   4 < 7   b   7 > 4   c   9 < 14

d   14 > 9   e   11 < 20   f   20 > 11

3   Hayley compared some numbers.
Did she use the correct signs?

a   3 < 5   b   24 > 20   c   48 > 91

d   59 < 48   e   100 > 50 + 50   f   80 + 1 < 90

➡ *Workbook page 15*

Order numbers

You can arrange numbers in order:

- from smallest to greatest

  [10] [20] [30]

- from greatest to smallest

  [95] [90] [85]

1 Is each set of numbers arranged smallest to greatest, or greatest to smallest?

a [9] [10] [11] [12]

b [27] [26] [25] [24]

c [60] [70] [80] [90]

d [38] [36] [34] [32]

2 Order the numbers from smallest to greatest.

a [18] [11] [10] [12] [15]

b [28] [19] [21] [25] [27]

c [81] [91] [51] [31] [61]

d [72] [98] [67] [53] [8]

💡 **Problem solving**

Work out the ages before you write them in order.

3 Raj is 2 years older than Ali.
Sarah is 3 years younger than Dev.
Dev is 1 year younger than Ali.
Raj is 9.
Write the names and ages in order from youngest to oldest.

➡ *Workbook page 16*

2D and 3D shapes

2D shapes

 **Think and share**

This kind of puzzle is called a tangram.

How many different **2D shapes** can you find in the tangram? Which shapes can you name?

Choose one of the shapes. Describe it without using its name.

1 Match the shapes to the names.

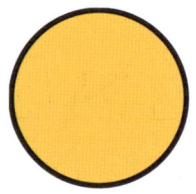 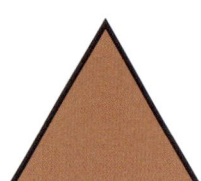  

| square | rectangle | circle | triangle |

2 Your teacher will give you some 2D shapes. Sort them into groups.

3 Draw some pictures using 2D shapes.

4 Look at your classmates' pictures. Name the different shapes.

Properties of shapes

How do you know which are triangles and which are squares?

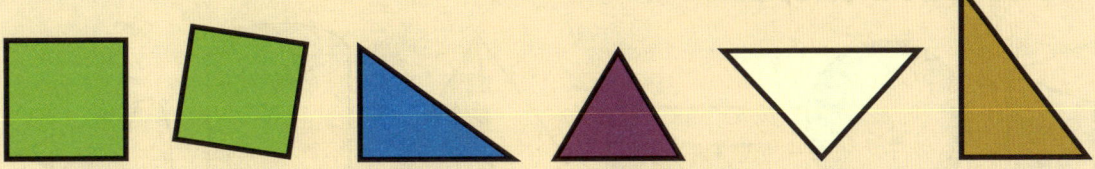

The **properties** of a shape tell us how to recognise that shape.
Properties of a 2D shape are things like:
- the number of **sides**
- the number of **vertices**.

Sides are straight lines.
A **vertex** is a point where the sides meet.
Vertices is the plural of vertex. We say 'one vertex, two vertices'.
Vertices are also called corners.

1 Match the descriptions to shapes you know.

a round shape with no vertices	has three sides and three vertices
has four equal sides and four vertices	has four **square corners** and four sides

2 Why are these shapes not rectangles?

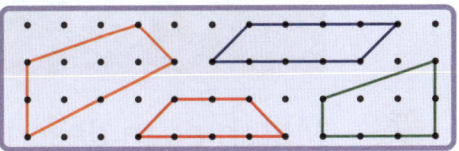

3 Maia made these shapes with elastic bands on a geoboard.

 a Which shapes are squares?

 b Which shapes are rectangles?

 c Which shapes are not squares or rectangles?

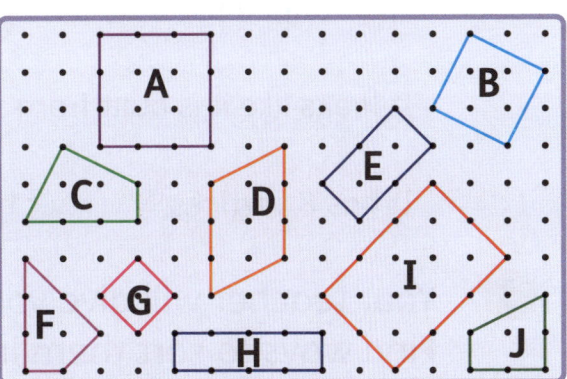

Identify 3D shapes

3D shapes are objects such as balls, boxes and cones.
Here are some 3D shapes:

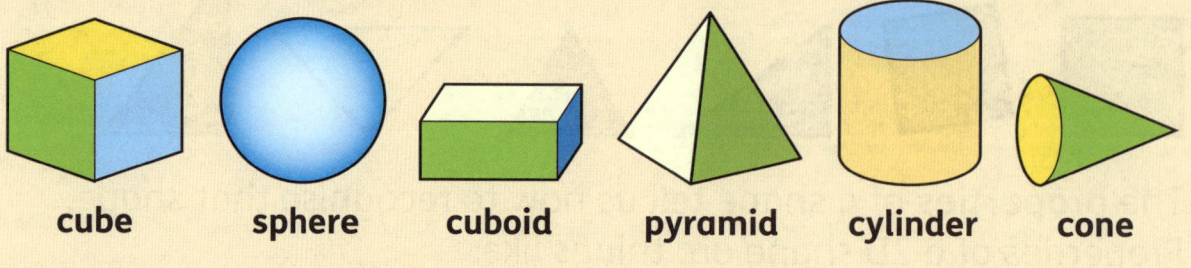

cube   sphere   cuboid   pyramid   cylinder   cone

In real life, many objects have these shapes.

1   Talk about the 3D shapes you can see in this picture.

2   Match each description to a 3D shape.

It is shaped like a ball.   It looks like a long box.

It looks like it is built from triangles.   It has no straight **edges**.

It has 8 vertices.   It has 5 vertices.   It has one vertex.

3   Your teacher will give you some 3D shapes.
Find ways to sort them into groups.

➡ *Workbook page 17*

Faces of 3D shapes

A **face** is a flat surface of a 3D shape.

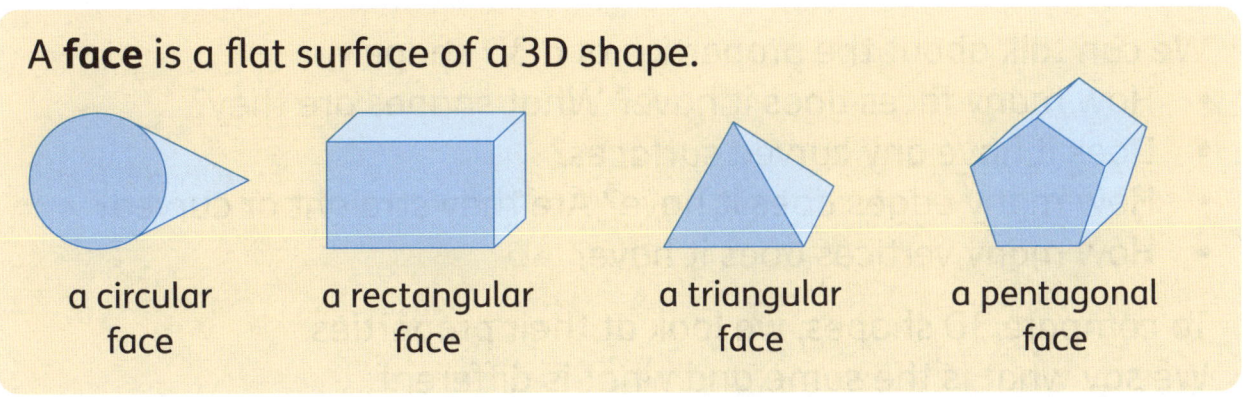

a circular face

a rectangular face

a triangular face

a pentagonal face

1 Which 2D shape can you find on all the shapes in each set?

a

b

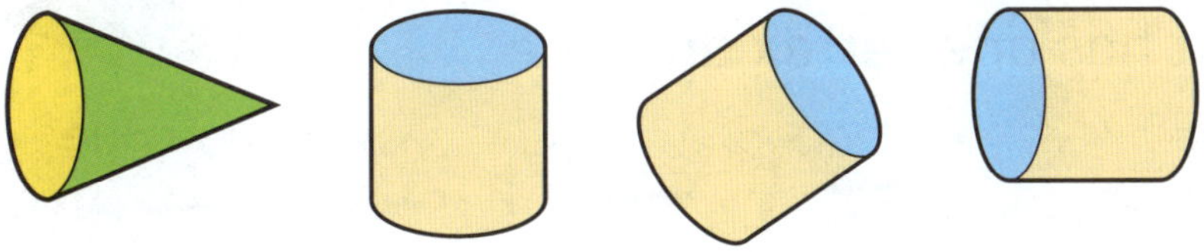

c

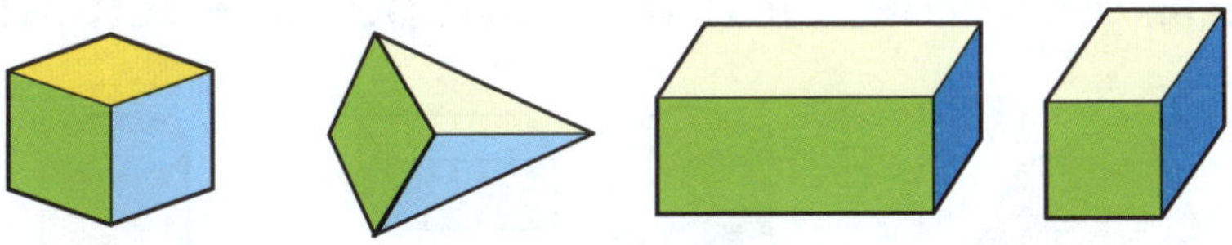

d

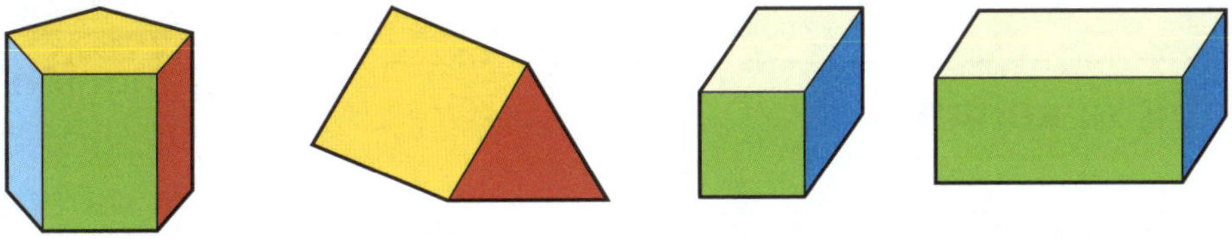

e

2 Find some 2D shapes on the faces of 3D shapes in your classroom.

Properties of 3D shapes

We can talk about the properties of a 3D shape.
- How many faces does it have? What shapes are they?
- Does it have any curved surfaces?
- How many edges does it have? Are they straight or curved?
- How many vertices does it have?

To compare 3D shapes, we look at their properties.
We say what is the same and what is different.

sphere

The sphere has no edges, vertices or faces.
It has only one curved surface.

cone

The cone has one curved surface and one circular face. It has one vertex and one curved edge.

1 Compare the properties of the 3D shapes in each pair.

a 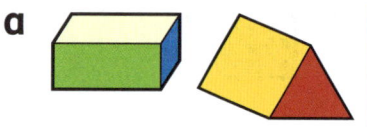 b   c

💡 **Problem solving**

Imagine you are looking directly at different faces of the 3D shape.

2 Each pair of shapes shows the same 3D shape from two different directions. Name the 3D shape. Explain how you worked it out.

a 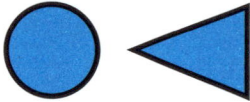 b  c

➡ *Workbook page 18 and page 19*

Circles

A circle is a closed curve with no vertices or sides.
The **centre** is the point in the middle of the circle.
All the points on the circle are the same distance from the centre.

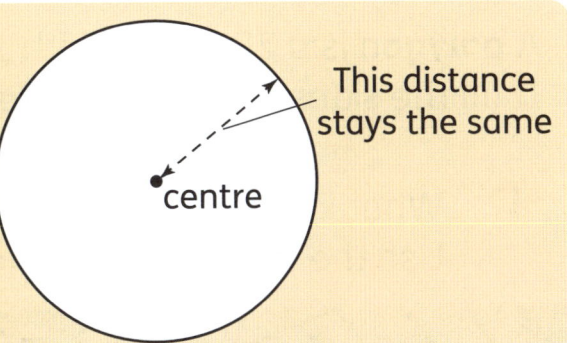

This distance stays the same

centre

1 Look at these different ways of drawing circles.
Why do they all work?
Find the centre of each circle.

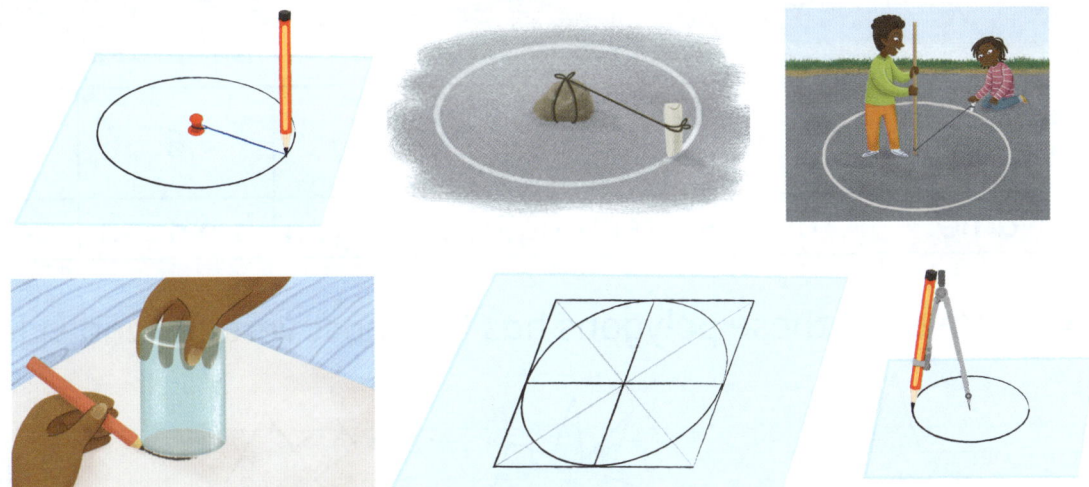

2 Here are some closed curves.
Why are they not circles?

3 Look at these circles.
How can you draw circles like this?

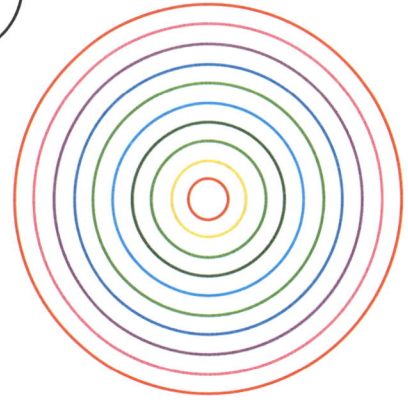

Polygons

A **polygon** is a 2D shape with straight sides.
If all the sides are the same **length**, we call it a **regular polygon**.

1. What kinds of polygons can you see in these pictures?
   Use the table below to help you.

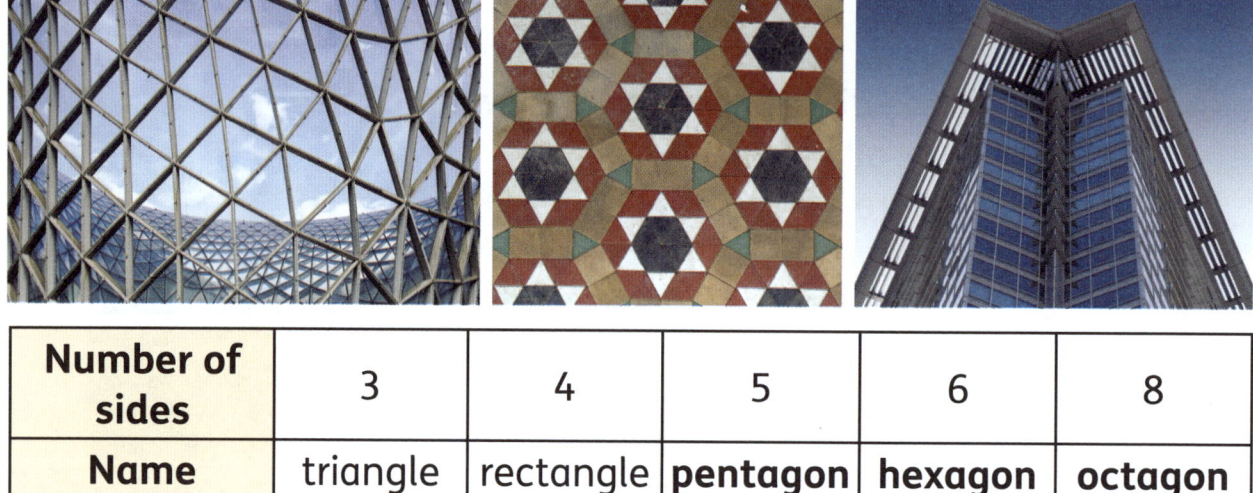

Number of sides	3	4	5	6	8
Name	triangle	rectangle	**pentagon**	**hexagon**	**octagon**

2. Which of these polygons has the most vertices and sides?

 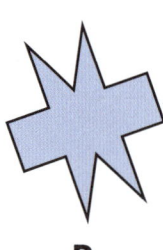  

A   B   C   D

 Problem solving

3. Can you make each of the shapes in the table in question 1 using only triangles?
   What is the smallest number of triangles you need to make each shape?

➡ *Workbook page 20*

Look at patterns

💭 Think and share

What do you notice about all these pictures?

Which real-life objects have **patterns** that people have made?
Which objects have natural patterns?
Which shapes do you notice?
Where do you see patterns in real life?

We can make patterns out of shapes, objects and numbers.

➡ *Workbook page 21*

Number   27

Repeating patterns

What is the next shape? How can you work it out?

This is a **repeating pattern**. The pattern repeats over and over.

and  are the **terms** of the pattern.
Terms are the things that repeat.

The **core** of the pattern is the set of terms that repeats.

The core of this pattern is  .

1 Find the terms and the core of each pattern. Say what shape comes next.

a

b

c

d

e

2 Find the terms and core of these number patterns.
Say which number should come next in the empty circle.

a (1) (2) (1) (2) (1) (2) (   )

b (3) (4) (5) (3) (4) (5) (   )

Find the rule

We can use letters to describe the terms, the core and the rule of repeating patterns, like this:

Rule:

AB

ABB

ABC

1   Look at the strings of beads.
    Work with a partner to decide whether the rule is
    AB, ABB or ABC.

a

b

c

d

2   Use the numbers 1 and 2.
    Arrange them in these patterns.

    a AB   **b** ABB   **c** ABA

➡️ *Workbook page 22*

Repeating shapes and numbers

Here are more patterns:

Rule:

ABA

ABBA

ABCA

1   Look at the shape patterns above.
Explain what the rules tell us about the terms and the
core of each pattern.

2   Use letters to write the rules of these number patterns.
Decide which term comes next.

a | 1 | 3 | 1 | 1 | 3 | 1 | 1 | |

b   (1)(2)(3)(1)(1)(2)(3)( )

c | 4 — 5 — 5 — 4 — 5 — 5 — |

d   [7 — 8 — 1 — 7 — 8 — 1 —]

e   (9)(4)(4)(9)(4)(4)( )

Problem solving

> There are many possible answers.

3   I am a pattern that follows the ABC rule.
All my terms are even numbers less
than 7. What pattern could I be?

Growing patterns

This is a **growing pattern**. It grows each time.

What are the next two shapes and numbers?
The rule is 'add two blocks each time'.

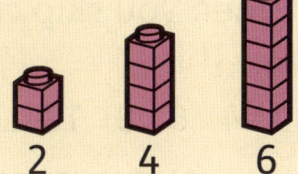

2   4   6

1   Explain the rule.
        Say what comes next in each pattern.

a

b

c

d

2   How do these number patterns grow?
        Say the rule and what comes next.

a 1   3   5   7   ☐

b 3   6   9   12   ☐

3   In some patterns the terms get smaller.
        Say the rule and what comes next.

a 10   8   6   4   ☐

b 17   16   15   14   ☐

4   How is a growing pattern different from a repeating
        pattern?

Skip-counting patterns

When you skip count, you count in the same group each time, like this:

2 4 6 8 10 ...

skip counting in groups of 2

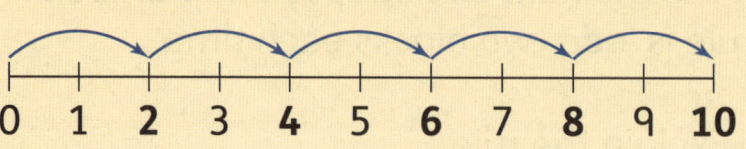

0 1 **2** 3 **4** 5 **6** 7 **8** 9 **10**

1 Skip count using each number line.

a In groups of 3

0 1 2 **3** 4 5 **6** 7 8 **9** 10 11 **12** 13 14 **15** 16 17 **18** 19 20 **21** 22 23 **24**

b In groups of 5

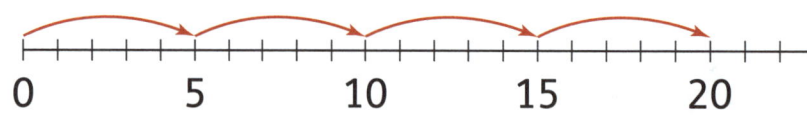

0   5   10   15   20

2 These number tracks show skip counting.
What is different about them?
What is the same?

0	10	20	30	40	50	60	70	80	90

0	5	10	15	20	25	30

💡 **Problem solving**

 Draw a number line or track to help you.

3 Shayna and Jack are skip counting in different ways. They both start at 0. Shayna's last two numbers are 28 and 32. Jack's last two numbers are 25 and 30.

a What size skips are they using?

b Which number will they both land on?

➡ *Workbook page 23 and page 24*

Add and subtract

Bar models

Think and share

What does this bar model show you?

5	3
8	

Sal made a stick of blocks.
He used it to make some addition
and subtraction sentences.

$$5 + 3 = 8 \quad 8 - 3 = 5$$
$$3 + 5 = 8 \quad 8 - 5 = 3$$

Talk about what Sal's stick and the bar model show.

1 Compare this bar model to the models above.
What do you notice?

15	3
18	

2 Write addition and subtraction sentences for this bar model.

2	7
9	

Problem solving

3 What number is
missing from this
bar model?

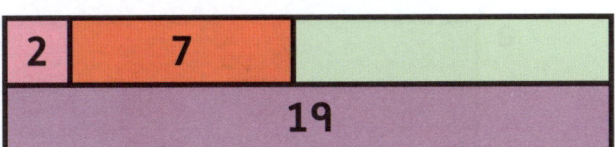

Add and subtract

What number is missing from this bar model?

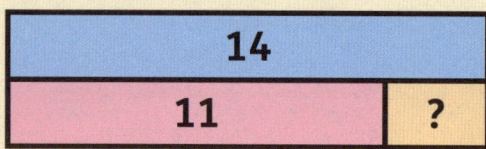

14
11 ?

Four pupils worked this out in different ways:

$11 + ? = 14$
$14 - 11 = ?$

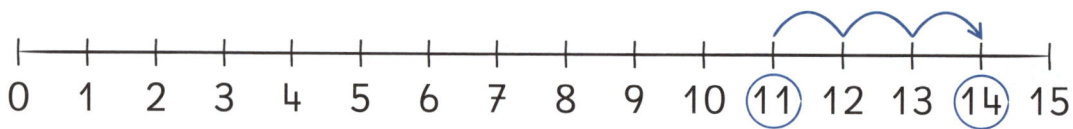

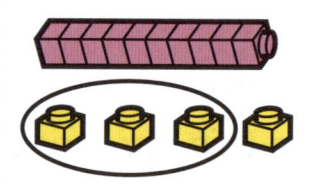

Describe the different ways they worked it out.

1 What number is missing from each bar model?

a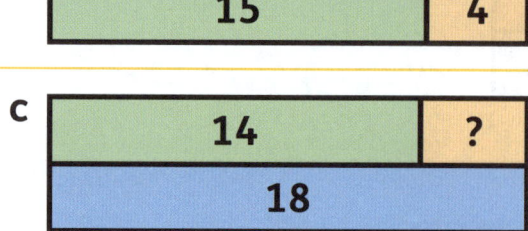

b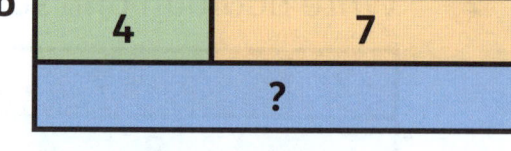

c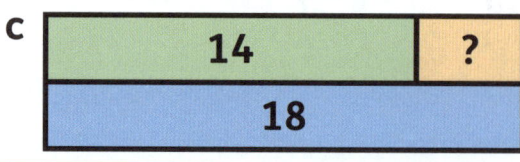

d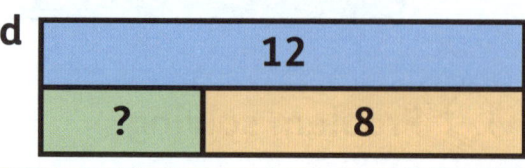

e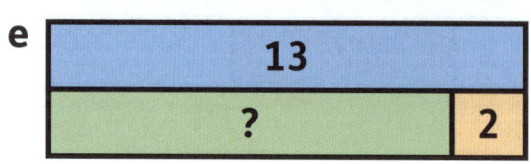

f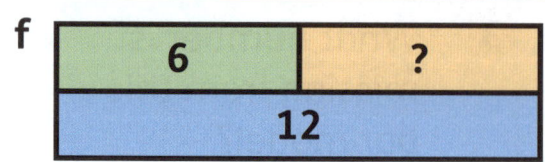

➡ *Workbook page 25*

Make 20

There are different ways to make 20.
We can count or skip count.
We can add. For example: $18 + 2 = 20$

1 **a** Count 20 bears.

   b How many groups of 10 make 20?

2 **a** Skip count in groups of 5.

   b How many groups of 5 make 20?

3 Complete the additions. They all equal 20.
   Use the pictures above or this number track to help you.

1	2	3	4	5	6	7	8	9	10	11	12	13	14	15	16	17	18	19	20

   a $20 = \boxed{} + 1$ **b** $\boxed{} + 5 = 20$

   c $10 + \boxed{} = 20$ **d** $20 = \boxed{} + 4$

   e $17 + \boxed{} = 20$ **f** $14 + \boxed{} = 20$

Add or subtract ones

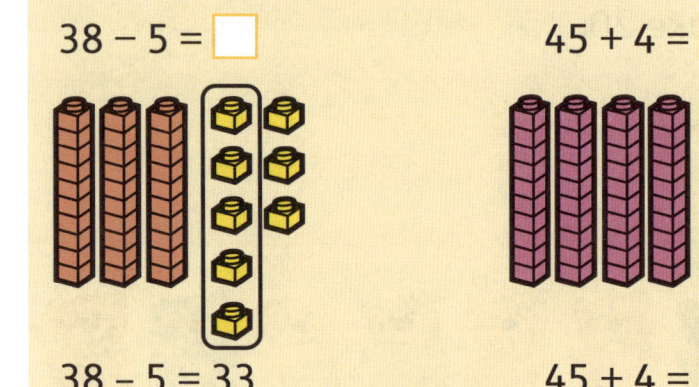

38 − 5 = ☐

38 − 5 = 33

45 + 4 = ☐

45 + 4 = 49

1 Say subtraction sentences to describe these pictures.
Subtract the circled ones.

a

b

c

2 Say addition sentences. Add the circled ones.

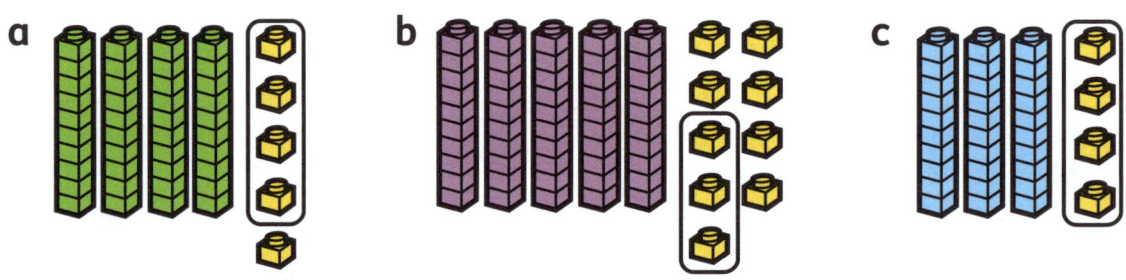

a

b

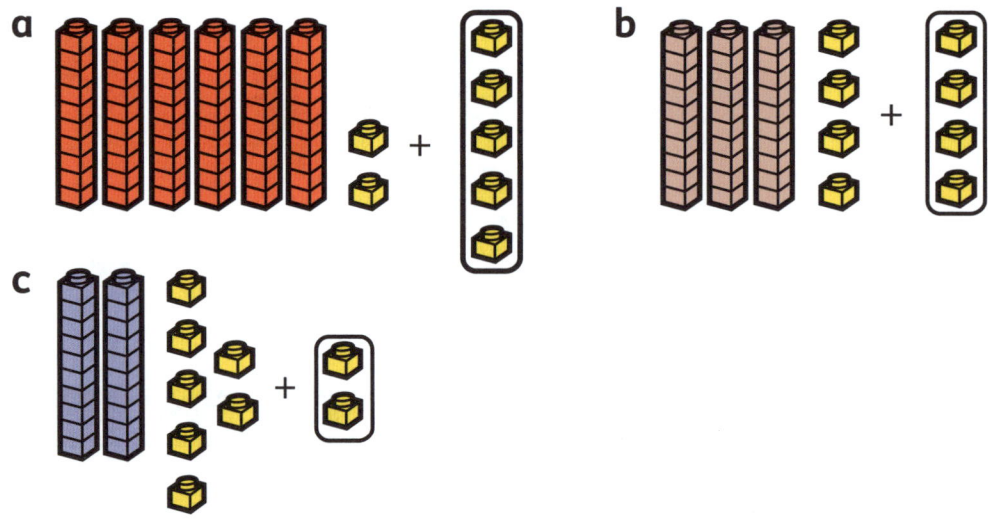

c

3 Work these out with your partner.

   a 89 − 5   b 75 + 3   c 43 − 2

   d 58 + 1   e 38 − 7   f 26 + 3

Work with tens

43 + 20 = ☐

These four pupils all worked it out differently.

Vikesh:

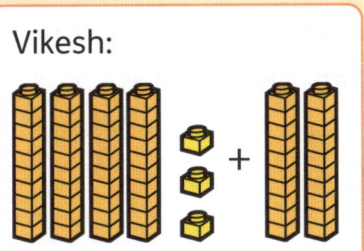

Nirman:

4 tens and 3 ones

plus

2 tens

Diya:

40	41	42	43	44	45	46	47	48	49
50	51	52	53	54	55	56	57	58	59
60	61	62	63	64	65	66	67	68	69

Sara:

Tens	Ones
4	3
2	0

1 **a** What did each pupil use to work out the answer?

    b What is the answer? Use their work to help you.

    c Which way of working do you find easiest to use? Why?

2 **a** Add 25 and 10.   **b** Find the total of 31 and 20.

    c What is 40 more than 55?   **d** What is 10 plus 17?

    e 60 + 15 = ☐   **f** 50 + 37 = ☐

3 **a** Start with 43. Take away 10. What is left?

    b Take 20 away from 65.

    c What is 94 minus 30?

    d 68 − 10 = ☐

➡ *Workbook page 26 and page 27*

Work with tens and ones

$38 - 25 = \square$

Tens	Ones
3	8
− 2	5
= 1	3

3 tens and 8 ones
take away: 2 tens and 5 ones
left: 1 ten and 3 ones

Isha checks her work like this: $13 + 25 = 38$ ✓

1 Use the way of working you like best to help you subtract.

> Make sure you understand the words **less** and **difference**. Your teacher can help you.

a Start with 49. Take away 17.

b What is 22 less than 53?

c Find the difference between 98 and 74.

2 Hannah says:

> 13 + 12 is the same as 12 + 13.
> Both make 25.
> That means 12 − 13 is the same as 13 − 12.

a What is Hannah's mistake?

b Why can you add in any order?

c Why can't you subtract in any order?

💡 Problem solving

> You can draw bar models and use place-value tables to help you.

3 The underlined number in each sentence is incorrect. First try to explain the mistake.
Then add or subtract to work out the correct answer.

a $75 - 14 = \underline{89}$ **b** $43 + 23 = \underline{26}$ **c** $27 - 13 = \underline{40}$

➡ *Workbook pages 28 to 30*

Add three numbers

$5 + 7 + 6 = \boxed{}$

Myra:

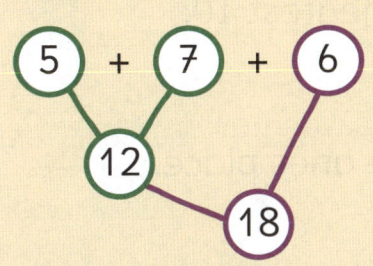

$5 + 7 = 12$
$12 + 6 = 18$

Shiv:

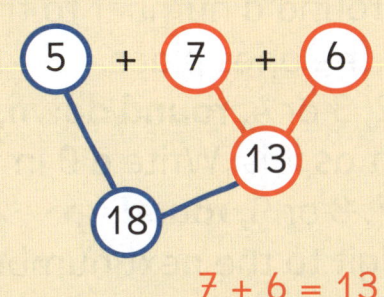

$7 + 6 = 13$
$13 + 5 = 18$

1 a What did Myra and Shiv do differently?

b Why did they both get the same answer?

c What other way could they have used?

2 Add three numbers.

a $2 + 3 + 7$ b $5 + 4 + 5$ c $1 + 8 + 4$

💡 Problem solving

3 Kendi has three plates of samosas.
How many samosas does Kendi have altogether?

4 Ola has three different plates of samosas.
The middle plate has 6 samosas.
One plate has 2 more than the middle
plate, and one plate has 3 fewer.
How many samosas does Ola have altogether?

> Draw a picture to help you.

➡ *Workbook page 31 and page 32*

Round to tens

Sometimes we **round** a number to the nearest 10 instead of giving the exact number.

Follow these steps to round a number to the nearest 10:
- Look at the digit in the ones place.
- If the digit is 0, 1, 2, 3 or 4, **round down**.
  Leave the tens digit as it is. Write a 0 in the ones place.
- If the digit is 5, 6, 7, 8 or 9, **round up**.
  Change the tens digit to the next number.
  Write a 0 in the ones place.

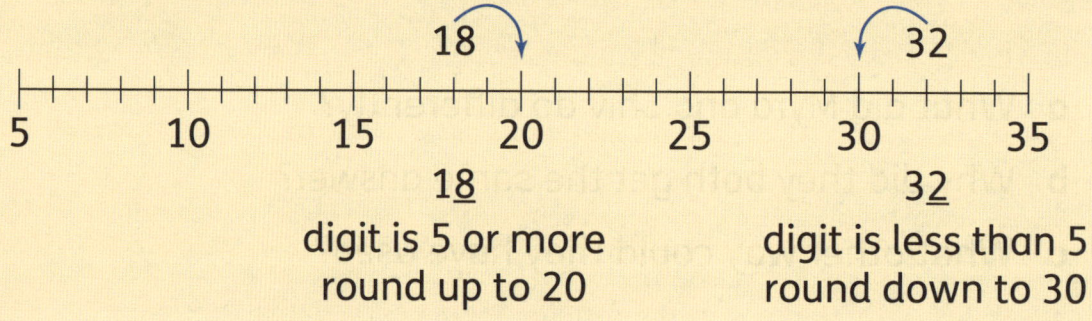

18
digit is 5 or more
round up to 20

32
digit is less than 5
round down to 30

1 Round each amount to the nearest 10.

    a 19 buttons rounds to ☐ buttons.

    b 52 pupils rounds to ☐ pupils.

    c 68 steps rounds to ☐ steps.

    d 82 pages rounds to ☐ pages.

2 Round these numbers to the nearest 10.

    a 47   **b** 91   **c** 85   **d** 34

3 Which of these numbers will round to 40?

37	41	45	51	49	44	33	35

Round to estimate

Rounding can help you to estimate.
First, round the numbers in the question.
Then work out an estimate for the answer.

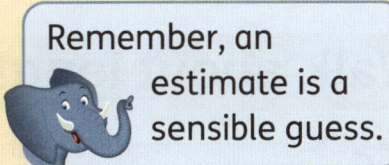

Remember, an estimate is a sensible guess.

For example:
23 pupils travel on Bus A and 38 pupils go on Bus B.
Estimate how many pupils go on the buses altogether.

To estimate, first round the numbers to the nearest 10.

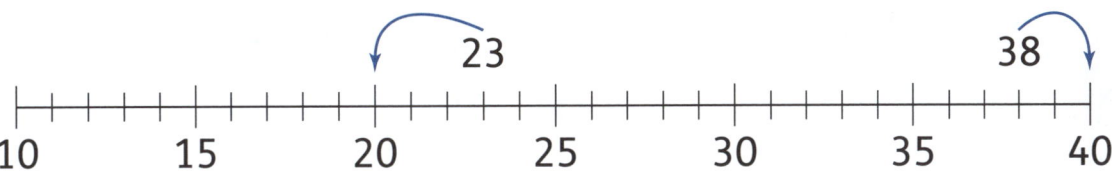

23 rounds down to 20. 38 rounds up to 40.
20 + 40 = 60
The estimate for the answer is 60.

Problem solving

First round the numbers. Then work out the estimate using tens.

1   Work out an estimate.

   a The spotty pencil case has 24 pens.
      The stripy case has 19.
      Estimate how many pens there are altogether.

   b There are 36 stickers on one sheet and 45 on another.
      Estimate how many stickers there are altogether.

   c Viraj has 89 stickers and gives away 43.
      Estimate how many stickers he has left.

UNIT 7 Length

Talk about length

Think and share

Length is a fixed distance between two points.
When we ask, 'How long is it?', we are asking, 'What is the length?'

Nila uses her finger to measure some lengths.

↔ 1 finger **width**

> I measured the length of my stapler. It is 8 finger widths.

Ethan uses a shoe.

1 shoe length

> My skateboard measures 4 times the length of my shoe.

How many finger widths is:
- your maths book
- your pencil
- your eraser?

1 Pranav used a short pencil to measure some everyday objects.

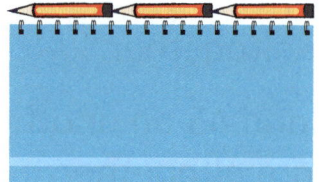

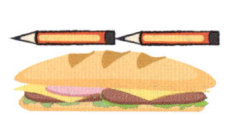

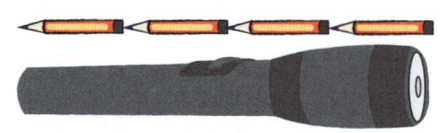

a Which object is longest?   b Which is shortest?

c How much longer is the torch than the notebook?

2 Why don't we usually use pencils to measure length?

Centimetres

Finger widths and shoes are **non-standard units**.
This means that they are not all the same.

When we want to measure a length exactly, we use
standard units like **centimetres** and **metres**.

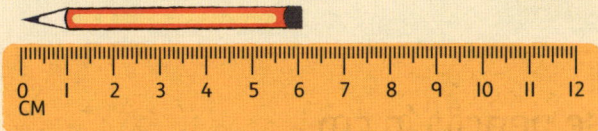

This **ruler** is marked in centimetres.
It shows a length of 12 cm.
The pencil is 6 cm long.

cm is a short way to write centimetres

1   Write the length of each pencil.
Use cm to show centimetres.

a
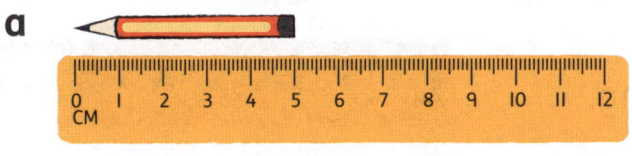

b
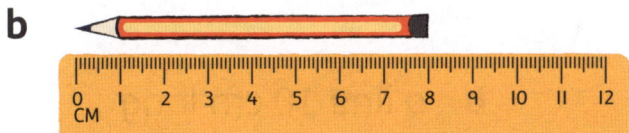

c
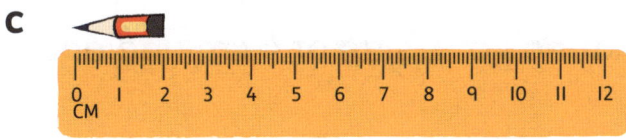

d

➡ *Workbook page 33 and page 34*

Measure and draw

When you measure using a ruler, always line up the object you are measuring with the 0 mark on the ruler.

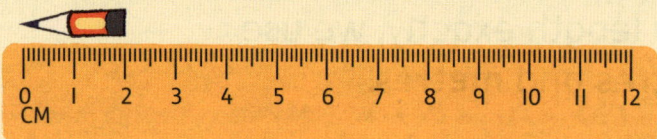

1 Measure the lengths of these pencils in cm.

a

b

c

d

e

f

2 Rian measured a paper clip like this.
He says it is 5 cm long.
Explain his mistake.

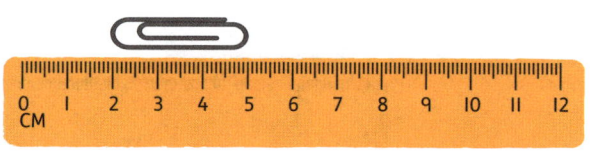

3 Draw:

    a a line 5 cm long   **b** a line 10 cm long

    c a square with sides that are 3 cm long.

4 **a** Draw a triangle that has one side of 4 cm and another side of 5 cm.

    b Measure the length of the third side.

    c Compare with a partner.
What looks the same, and what looks different about your triangles?

➡ *Workbook page 35 and page 36*

The metre

In real life, a **metre stick** is exactly 1 metre long.

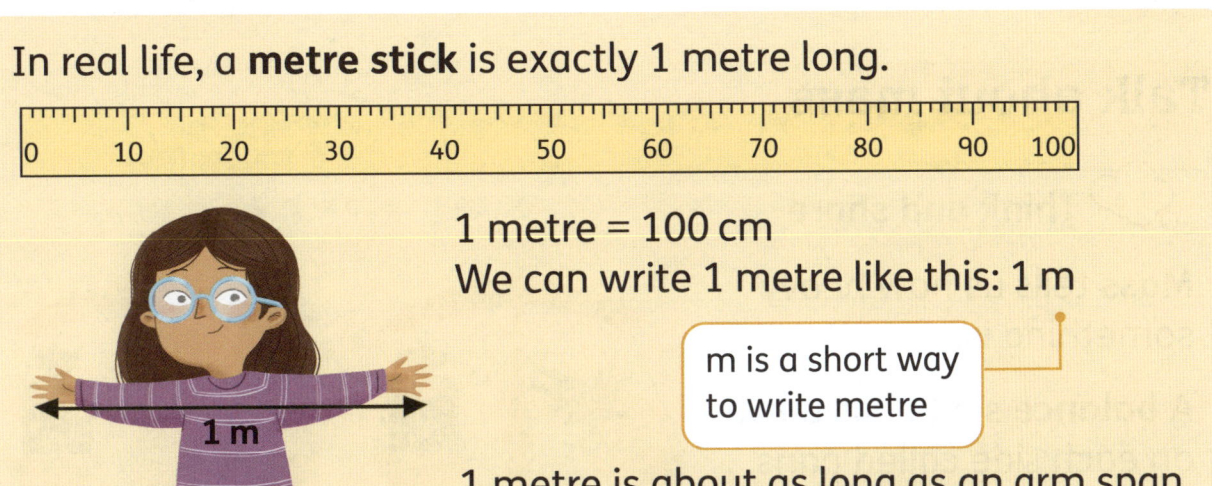

1 metre = 100 cm
We can write 1 metre like this: 1 m

m is a short way
to write metre

1 metre is about as long as an arm span.

1 Is each object: shorter than 1 metre, about 1 metre long,
or longer than 1 metre?
Write < 1 m, = about 1 m, or > 1 m.

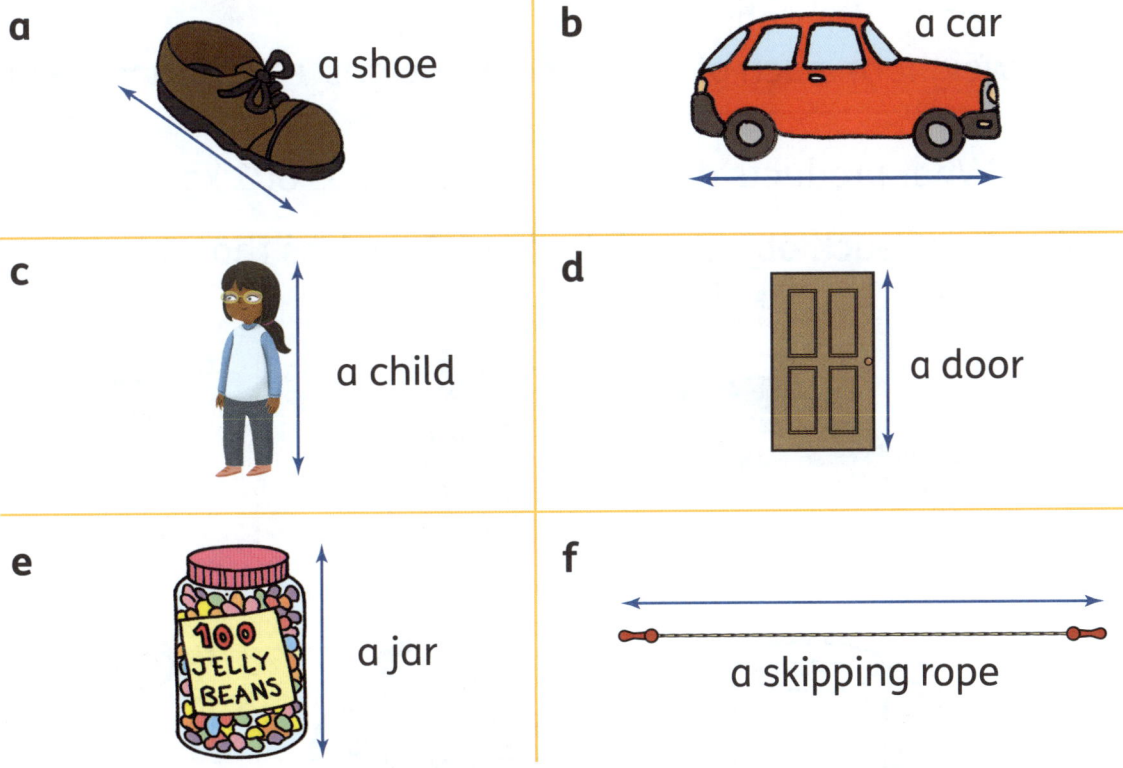

a   a shoe

b   a car

c   a child

d   a door

e   a jar

f   a skipping rope

2 Use a metre stick.
Find three things at school that are about 1 metre long.

Talk about mass

💭 Think and share

Mass tells us how heavy something is.

A **balance scale** has bowls on each side called pans. When the pans are level, it shows the objects on both sides are **equal** in mass. We say the scales are balanced.

We can use a **weight** to measure how heavy other objects are. The weight on each of these scales weighs 1 **kilogram** (1 kg).

1 What products do we buy in packages of 1 kg?

2 Does each object weigh about 1 kg, less than 1 kg or more than 1 kg?

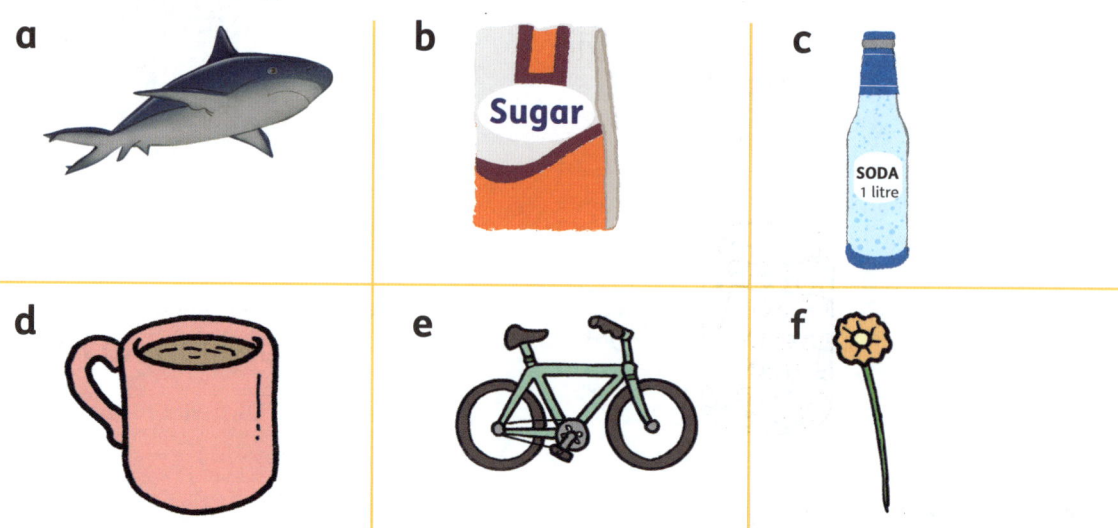

a b Sugar c SODA 1 litre

d e f

➡ *Workbook page 37 and page 38*

Compare mass of objects

We can use the < and > signs to compare masses.

The big sack of potatoes is heavier than the small bag.

10 kg > 1 kg

1 Look at these sea creatures.

starfish

lobster

dolphin

shark

triggerfish

seahorse

a Which is the heaviest?

b Which is lighter, the seahorse or the lobster?

c Write the names of the creatures in order from lightest to heaviest.

d Next to the name of each creature, write whether you think it has a mass of more than 1 kilogram (> 1 kg), about 1 kg, or less than 1 kilogram (< 1 kg).

We weigh light objects in units called **grams**.
A gram is about the mass of two paper clips.

2 Use a balance scale to find out the mass of:

a a pencil **b** an eraser **c** a ruler.

➡ *Workbook page 39 and page 40*

Solve problems

When you solve problems, make sure that all the objects are measured in the same units.

On this page, all the masses are in kg.

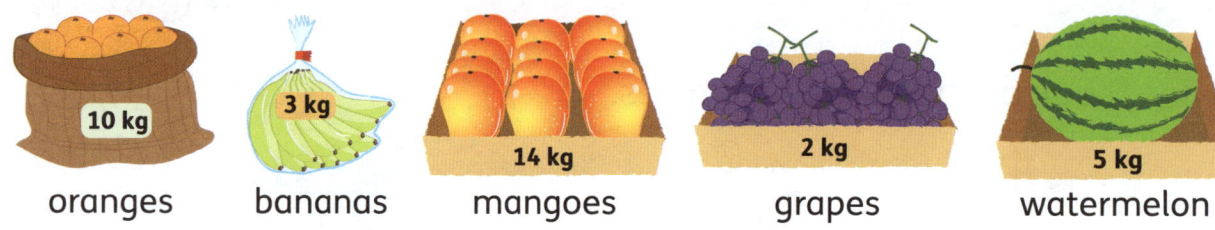

| oranges | bananas | mangoes | grapes | watermelon |

1 Look at the pictures.

    a Which items are heavier than the watermelon?

    b Which item is lighter than the bananas?

    c How much heavier are the oranges than the watermelon?

    d How much lighter are the bananas than the mangoes?

2 Work out the mass of:

    a oranges + watermelon   **b** mangoes + oranges

    c mangoes + bananas + grapes.

Problem solving

3   **a** What can you say about the watermelon and the weights?

    b If the watermelon weighs 4 kg, what can we say about the weights?

    c If the watermelon weighs 2 kg, what can we say about the weights?

    d If each weight is half a kilogram, what can we say about the watermelon?

Mixed practice 1

1 Teia showed the number 6 like this, using counters. Draw two different = ways you can arrange counters to show the number 6.

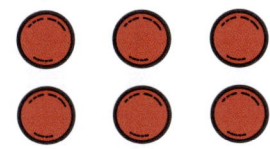

2 The first tin contains beans.

    a Which tins are second, third and fourth in the row?

    b In which positions are the coconut milk and the corn?

    c Write two more sentences about the positions of the other tins.

3 Write the number names for these numbers.
Then write how many tens and how many ones.

    a 48   **b** 92   **c** 80   **d** 39   **e** 25

4 Write these numbers in order from smallest to greatest.

    a 11, 18, 4, 23   **b** 53, 32, 23, 35

5 What 2D shape am I?
Draw the shape and write what it is called.

    a I am a round shape with no corners.

    b I have four square corners, but my sides are not all equal.

6 Write the name of each 3D shape.

    a 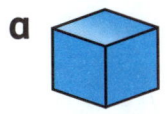   **b**    **c** 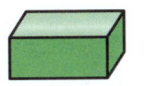   **d**    **e**    **f**

7 Draw or write your own repeating pattern using three different shapes. Explain how your pattern repeats.

8 Skip count in groups of 5 to 50. Write all the numbers.

9 Measure the height of each tower.
Then complete the statements using the sign < or >.

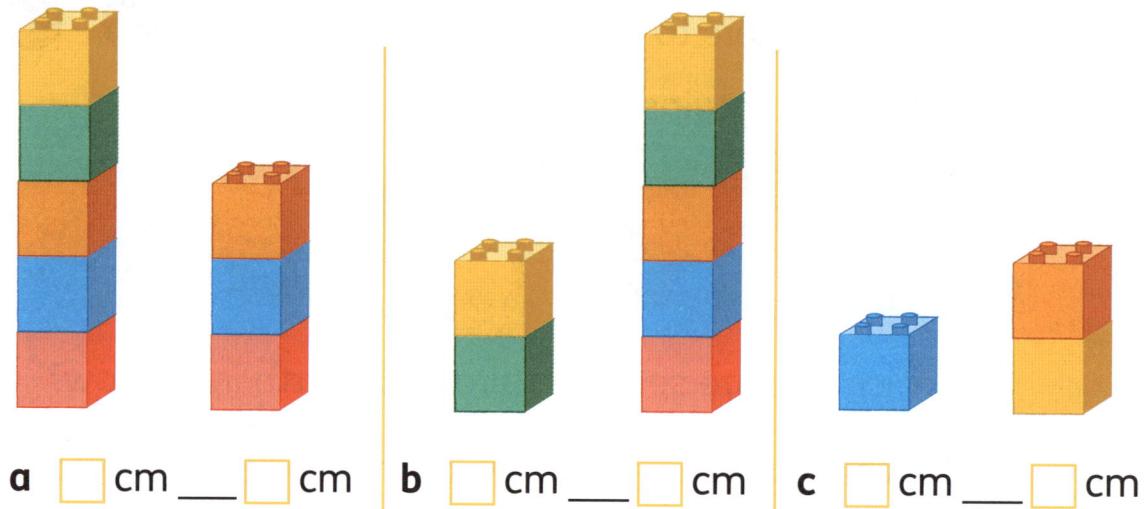

a ☐ cm ___ ☐ cm   **b** ☐ cm ___ ☐ cm   **c** ☐ cm ___ ☐ cm

10 A sunflower was 15 cm tall.
It grew another 10 cm.
What is the new height?

> You can use the different ways of working you learnt in Unit 6.

11 Mike built a tower.
Then he made it 10 cm taller.
Now the tower is 35 cm.
What was the height before Mike made it taller?

12 Sara built a tower 17 cm taller than Kevin's tower.

  a What do you need to know to work out how tall Sara's tower is?

  b Sara's tower is 29 cm tall. How tall is Kevin's tower?

13 For each scale, say whether the fruit or vegetable is more than, less than or equal to 1 kg.

Lists and tables

Ask questions

 **Think and share**

Sam sells many different flavours of ice cream in his shop.

V dairy-free

How many different flavours are there?

How many sorbets are there?

How many fruit flavours are there?

Sam says he has two kinds of ice cream in his shop.

What do you think he means?

Suggest two other ways to sort the ice cream.

 Problem solving

1 Sam asked people to vote on new ice cream flavours.

   a How many people chose each flavour?

   b Which new flavour should Sam make?

Lime	ЖЖ ЖЖ
Honey	ЖЖ III
Cherry	ЖЖ II

➡ *Workbook page 41*

Draw tally marks

You have seen the **tally table** that Sam drew on page 51.
He used **tally marks** to show how many people chose each flavour.

| = 1

卌 = 5

1 Draw tally marks for these numbers.

    a 1   **b** 3   **c** 5   **d** 10   **e** 14   **f** 19

2 <, > or =?

    a |||| ____ 5   **b** |||| ____ 4   **c** 10 ____ 卌 ||||

    d 卌 | ____ 6   **e** 卌 卌 || ____ 9   **f** 卌 | + 卌 | ____ 20

3 Copy the tables below. Draw tally marks to show how many of each kind of ice cream Sam had in his shop.
Use the picture on page 51 to help you.

Fruit	
Not fruit	

Sorbet	
Not sorbet	

White		
Yellow		
Pink		
Brown		
Green		

➡ *Workbook page 42*

Venn diagrams

A **Venn diagram** uses circles to sort things into different **categories**. These two Venn diagrams sort polygons in different ways.

1. **a** How many shapes can you see in each diagram?

   b Look at the first diagram. What can you say about the shapes that are outside the circle?

   c Look at the second diagram. What can you say about the shapes that are in each circle?

   d How are the shapes sorted differently in the two diagrams?

2. Sally sorted a set of 3D shapes like this:

   a How many have *only* flat faces?

   b How many have curved surfaces?

   c How many shapes did Sally sort altogether?

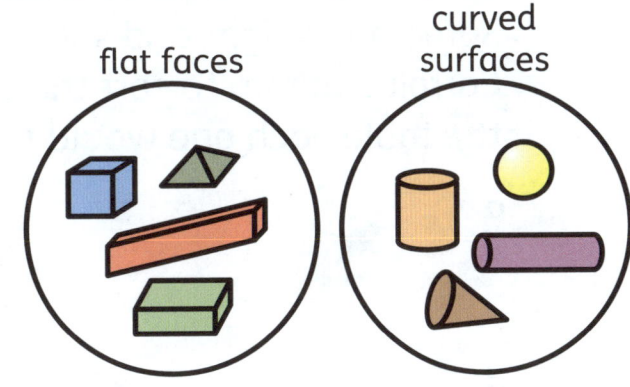

3. Riedwan sorted his shapes like this. Explain what he did

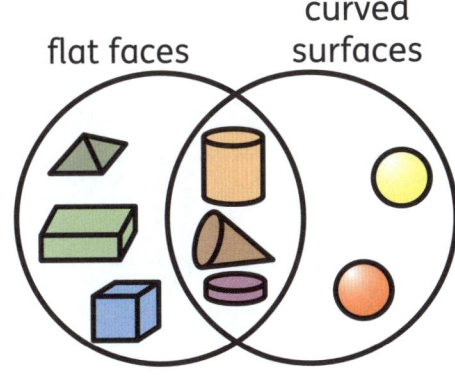

Carroll diagrams

You have used Venn diagrams to sort things into categories.
We can also sort things in tables.
This **Carroll diagram** has 2 **columns** and 2 **rows**.

	Grows on a plant	Does not grow on a plant
Food		
Not food		

1 Look at these objects.
Say which part of the table each one would go in.

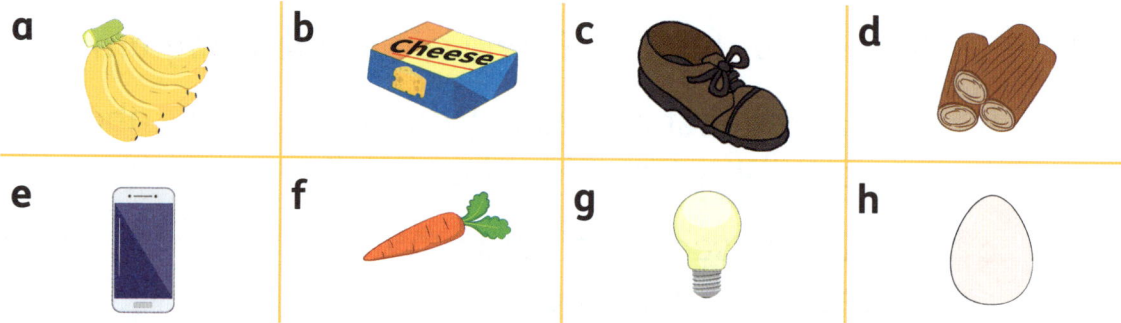

2 Some of these objects use batteries and some do not.
Some make light and some do not. Draw your own
Carroll diagram to sort the objects. Say which part of
the table each one would fit into.

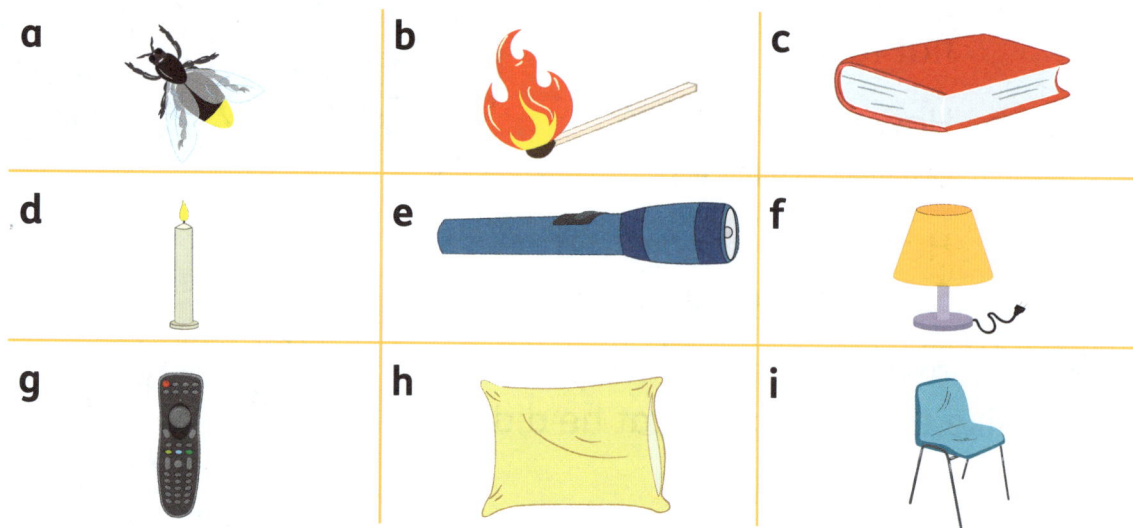

➡ *Workbook page 43 and page 44*

Sort data

The information we put in tables and diagrams is called **data**.
You can collect your own data and sort it into tables.

1 Do you recognise all the vegetables in this picture?
Name as many as you can.

2 Talk about the fruit and vegetables your family eats.
Make a list.
Make a poster with pictures of all the different fruit and vegetables.

3 Sort the fruit and vegetables in different ways. Draw diagrams like these to help you.

	Big	**Small**
Fruit		
Vegetable		

peel it   don't peel it

eat raw   eat cooked

UNIT 10 Show data

Represent data

 **Think and share**

A Year 2 class made this birthday chart.

Birthday chart

April	Grace 4th			
May	Andrew 1st	Diya 2nd	Pranit 8th	Zion 8th
June	Olivia 3rd	Mira 19th		

What does the chart show?

How is the data organised?

How could you show this data on a **pictogram** or **block diagram**?

How many pupils have birthdays in May?

How many in June?

Which month has fewer birthdays, April or May?

1 Grace collects the data from the birthday chart in a tally table.

    a Is Grace's data correct?

    b What information from the table has she left out?

Month	Number of birthdays
April	I
May	IIII
June	II

Pictograms

A pictogram uses small pictures or symbols in a table.
A pictogram always has a **key** to tell you what each picture or symbol represents.

1
   a Why do you think this kind of chart is called a pictogram?

   b What does 1 medal show?

   c Who won the most races?

   d Who won the fewest races?

   e Does the chart tell us how many races each runner took part in?

Races won

Kiara	🏅🏅🏅🏅🏅🏅🏅🏅🏅
Gianna	🏅🏅🏅🏅🏅🏅
Fatima	🏅🏅🏅🏅🏅
Navva	🏅🏅

Key: 🏅 means 1 race won

💡 Problem solving

2
   a The title of this pictogram is missing. What do you think the title might be?

   b The pictogram is not finished. It doesn't show how many pupils chose orange juice. A total of 25 pupils chose a juice flavour. How many pupils chose orange?

Juice flavour	Number of pupils
Mango	👤👤👤👤👤👤👤
Pineapple	👤👤👤👤👤👤👤👤👤
Apple	👤👤👤👤
Orange	

Key: 👤 = 1 pupil

Think about how many pupils chose each of the other flavours.

More pictograms

We can use a pictogram to help sort data into categories.
We put each category on one row.

1 What are the categories of juice flavours on page 57?

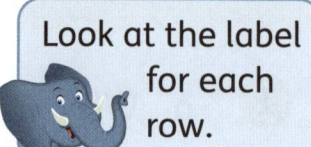

Look at the label for each row.

2 A jar of beads has four different shapes.

Complete the sentences about the beads.

Shapes of beads in a jar

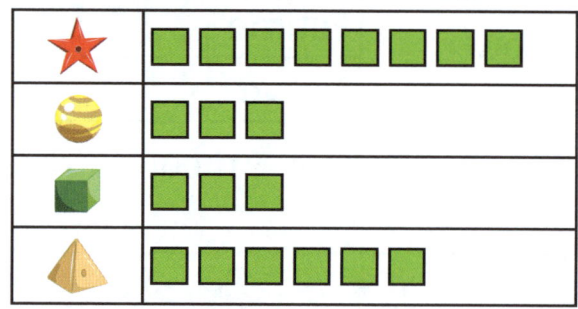

Key:
☐ = 1 bead

a The most common shape is the ___.

b There are equal numbers of ___ and ___.

c There are ☐ more pyramids than cubes.

d There are ☐ beads altogether.

💡 **Problem solving**

3 Draw a pattern for a necklace that uses all the beads.

---

4 Draw your own pictogram to show the different kinds of sweets.

➡ *Workbook page 45*

Block diagrams

A block diagram shows data using rows of blocks.
Always remember to label the blocks with the numbers they represent.

1 Complete the sentences about the block diagram.

Favourite snack in Joti's class

Chips, Rotis, Spring rolls, Samosas

1 2 3 4 5 6 7
Number of pupils

  a The snack most pupils like is ___.

  b More pupils like spring rolls than ___.

  c ☐ pupils chose spring rolls.

  d Equal numbers of pupils chose ___ and ___.

2 How can you work out how many pupils Joti asked?

3 How will Joti's block diagram help him to plan what snacks to make for his classmates?

4 Isabel asked some classmates which device they use to watch videos online.

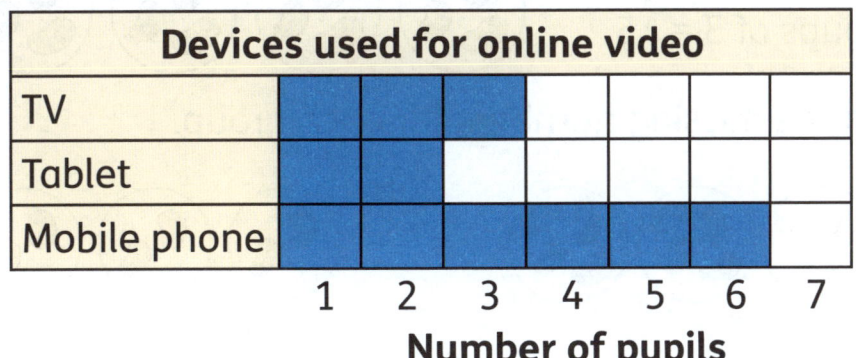

Devices used for online video

TV, Tablet, Mobile phone

1 2 3 4 5 6 7
Number of pupils

Make up three questions to ask a partner about Isabel's block diagram.

➡ *Workbook page 46 and page 47*

11 Multiply

Count in groups

Think and share

Look at this checkers board.

How many different shapes can you see?
How many beads are in each triangle?
How can you work out how many beads there are on the board, without counting all of them?

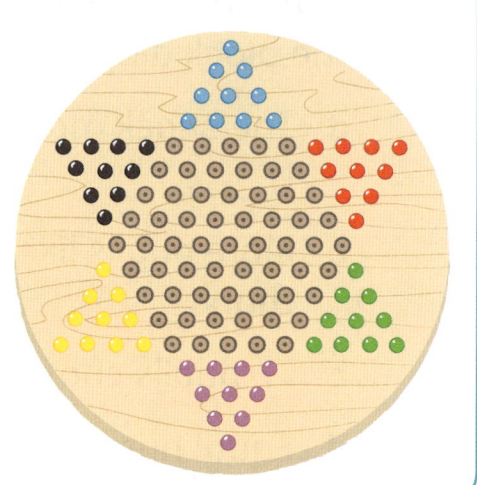

1. Look at this pattern made from pencils. How many pencils would you need for:

   a 3 triangles **b** 4 triangles **c** 5 triangles?

2. Here are 5 groups. There are 3 beans in each group.

   3 + 3 + 3 + 3 + 3 = 15
   5 groups of 3 = 15

   Fill in the missing numbers for each group.

   a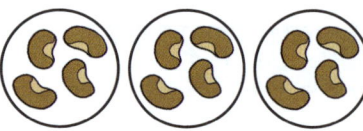

   ☐ groups of ☐

   ☐ + ☐ + ☐ = ☐

   b

   ☐ groups of ☐

   ☐ + ☐ + ☐ + ☐ = ☐

➡ *Workbook page 48*

Repeated addition

On the checkers board there are 6 groups of 10 beads.

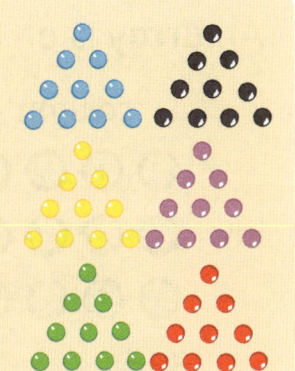

$10 + 10 + 10 + 10 + 10 + 10 = 60$
There are 6 tens.
We say 'six times ten'. The sign for **times** is ✕.
We write $6 × 10 = 60$

Another word for times is **multiply**.
Multiply means to repeat something a number of times.

1 Count the groups. Work out how many altogether.

a
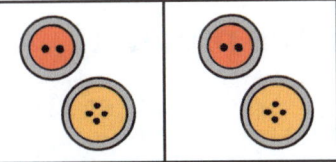

☐ groups of ☐
☐ + ☐ = ☐

b
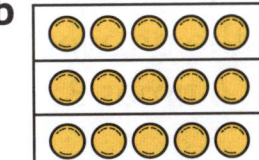

☐ groups of ☐
☐ + ☐ + ☐ = ☐

c

☐ groups of ☐
☐ + ☐ + ☐ + ☐ = ☐

d

☐ groups of ☐
☐ + ☐ + ☐ + ☐ = ☐

e

☐ groups of ☐
☐ + ☐ + ☐ = ☐

f

☐ groups of ☐
☐ + ☐ + ☐ + ☐ + ☐ + ☐ = ☐

➡ *Workbook page 49 and page 50*

Arrays

An **array** is an arrangement of objects like this:

column

row ○○○○
○○○○
○○○○

3 rows
4 dots in each row
3 × 4 = 12

We can also say:
4 columns
3 dots in each column
4 × 3 = 12

Repeat adding and arrays are both ways to multiply numbers.
When we multiply, the answer is called the **product**.
12 is the product of 4 times 3.

1 How many rows of objects can you see in each array?
How many objects in each row? Write multiplication sentences
for each picture. Some pictures have two multiplication
sentences, but some have only one.

a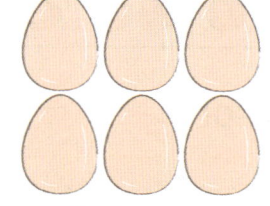

b

c

d

e

f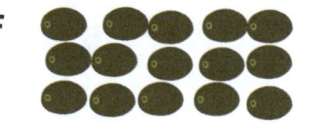

2 The example above shows you that 4 × 3 is the same as 3 × 4.
How do the arrays on this page show you that you can
multiply numbers in any order to get the same product?

➡ *Workbook page 51*

Multiply by 0 or 1

1 group of 3 apples		$1 \times 3 = 3$
3 groups of 1 apple		$3 \times 1 = 3$
0 groups of 3 apples		$0 \times 3 = 0$
3 groups of 0 apples		$3 \times 0 = 0$

1 **a** What do you notice about multiplying by 1?

   b What do you notice about multiplying by 0?

2 **a**

1 plate of 4 pears

$1 \times 4 = \boxed{}$

b

4 plates of 1 pear

$4 \times 1 = \boxed{}$

c

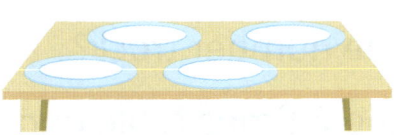

4 plates of 0 pears

$4 \times 0 = \boxed{}$

d

0 plates of 4 pears

$0 \times 4 = \boxed{}$

3 Say the missing number.

a $5 \times \boxed{} = 0$ **b** $0 \times 5 = \boxed{}$ **c** $1 \times \boxed{} = 6$

d $5 \times \boxed{} = 5$ **e** $8 \times \boxed{} = 0$ **f** $0 \times 10 = \boxed{}$

The 2 times table

$1 \times 2 = 2$

$2 \times 2 = 4$

$3 \times 2 = 6$

$4 \times 2 = 8$

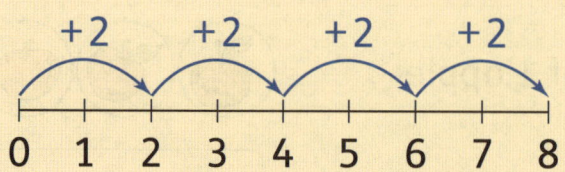

1	2	3	4	5	6	7	8	9	10
11	12	13	14	15	16	17	18	19	20

1 This is a number frame with 2 holes:

   a 1 number frame shows 1 group of 2. $1 \times 2 = \boxed{}$

   b 2 number frames show $\boxed{} \times \boxed{} = \boxed{}$

   c 3 number frames show $\boxed{} \times \boxed{} = \boxed{}$

2 **a** What do you think we mean by 'the 2 **times table**'?

Explain your ideas to a partner.

   b What do you notice about the number track?

   c What do you notice about the 2 times table and even numbers?

3 Multiply.

   a $5 \times 2 = \boxed{}$ **b** $6 \times 2 = \boxed{}$ **c** $7 \times 2 = \boxed{}$

   d $2 \times 5 = \boxed{}$ **e** $2 \times 6 = \boxed{}$ **f** $2 \times 7 = \boxed{}$

➡ *Workbook page 52 and page 53*

Multiply by 5 or 10

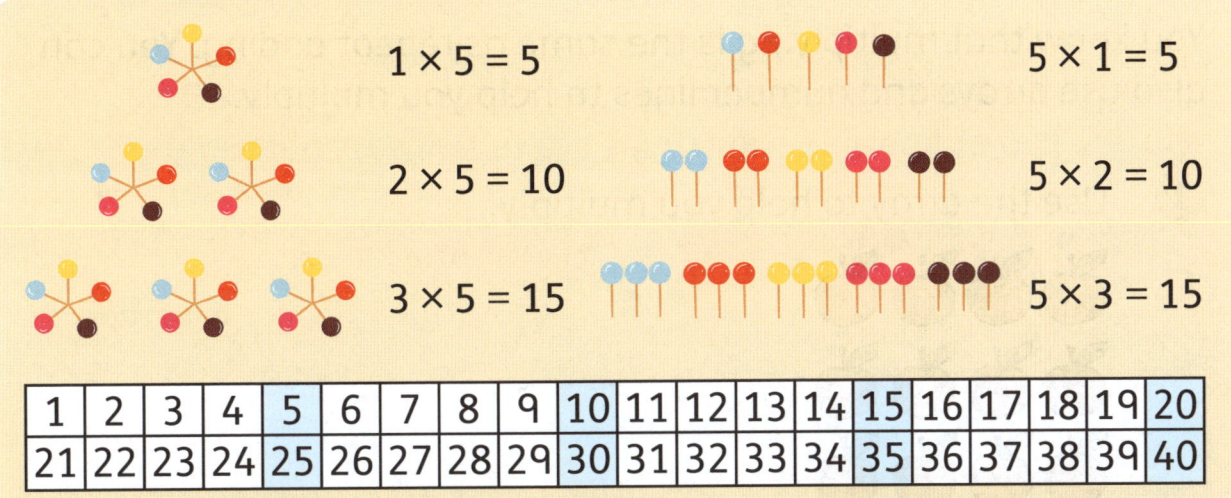

$1 \times 5 = 5$ $5 \times 1 = 5$

$2 \times 5 = 10$ $5 \times 2 = 10$

$3 \times 5 = 15$ $5 \times 3 = 15$

1	2	3	4	5	6	7	8	9	10	11	12	13	14	15	16	17	18	19	20
21	22	23	24	25	26	27	28	29	30	31	32	33	34	35	36	37	38	39	40

How do the pictures and the number track show the 5 times table?
Explain your ideas to a partner.

1 Use the number track to help you skip count in groups of 5.
Then carry on up to 100.

2 Look at this number line.

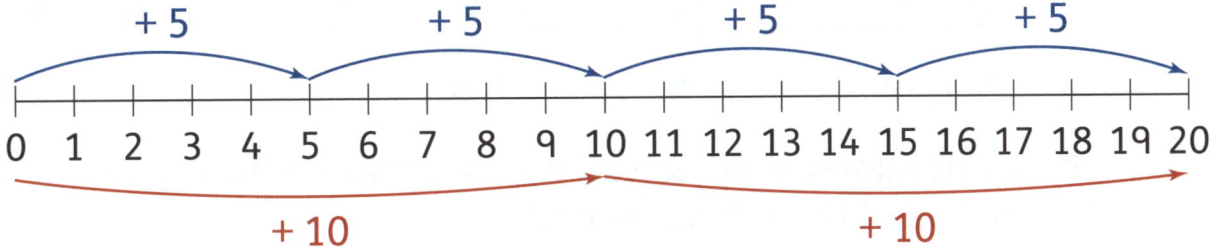

What do you notice about skip counting in groups of 5
and skip counting in groups of 10?

3 Use counters or objects to help you multiply.

a $20 = \boxed{} \times 5$ **b** $30 = \boxed{} \times 5$ **c** $40 = \boxed{} \times 5$

  $20 = \boxed{} \times 10$ $30 = \boxed{} \times 10$ $40 = \boxed{} \times 10$

4 Do you notice a pattern when you multiply by 5 or 10?
Tell your ideas to a partner.

➡ *Workbook page 54*

Multiplication practice

You know that multiplying is the same as repeat adding. You can also use arrays and number lines to help you multiply.

1 Use the array to help you multiply.

a $1 \times 12 = \square$

b $3 \times 4 = \square$

c $2 \times \square = 12$

d $\square \times 3 = 12$

e $0 \times 12 = \square$

2 Use this number track to help you answer the questions.

4	8	12	16	20	24	28	32	36	40
44	48	52	56	60	64	68	72	76	80

a $2 \times 4 = 8 \times \square$

b $4 \times 4 = 8 \times \square$

c What relationship do you notice between the 4 times table and the 8 times table?

d Which numbers belong to the 4 times table and the 10 times table?

e Find numbers from the 3 times table. What pattern do you notice?

Try counting in groups of 5 forwards and backwards, or making an array.

💡 **Problem solving**

3 A packet contains 5 shapes.
Preetal says: 'I used three packets, and now I have 10 shapes left.' How many packets did she start with?

Solve multiplication problems

You know how to multiply in different ways:

- skip counting
- number lines
- adding groups
- remembering patterns.

You can use any of these ways to solve problems.

Problem solving

1 In a game, each player has the same number of shapes.

    a How many players are there?

    b How many shapes does each player have?

    c Write the multiplication: ☐ × ☐ = ☐

2 How many shapes are on each table?

a

2 players
3 shapes each

b

3 players
5 shapes each

c

4 players
3 shapes each

d

6 players
2 shapes each

➡ *Workbook page 55*

Repeated subtraction

Think and share

Jaime has 12 samosas.
She wants to put 3 on each plate.
How many plates does she need?

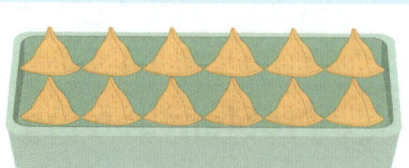

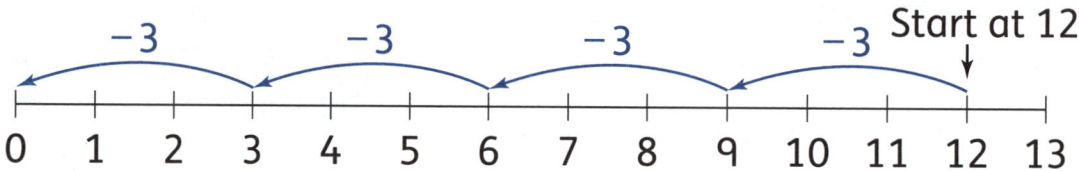

Start at 12

Jaime divided 12 into groups of 3.
The sign for **divide** is ÷.
$12 ÷ 3 = 4$

1 Divide.

a

8 sandwiches. 2 on each plate.
How many plates?
$8 ÷ 2 = \square$

b

10 cakes. 5 on each plate.
How many plates?
$10 ÷ 5 = \square$

c

20 wheels. 4 on each car.
How many cars?
$20 ÷ 4 = \square$

d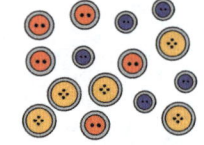

15 buttons. 3 on each shirt.
How many shirts?
$15 ÷ 3 = \square$

➡ *Workbook page 56*

Equal sharing

I want to share 15 dumplings equally between 5 plates.
How many should there be on each plate?

$15 \div 5 = ?$

I put one dumpling on each plate.
Then another one on each plate.
I continue until there are none left.

I know that $3 \times 5 = 15$.
So, if I want to divide 15 into 5 equal groups, each group will have 3.

$15 \div 5 = 3$

1 Why do both ways of working out get the same answer?

2 Share equally.

a 12 bananas
3 bowls
$12 \div 3 = \square$

b 3 hooks
3 bags
$3 \div 3 = \square$

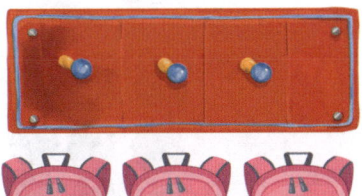

c 16 balls
4 drawers
$16 \div 4 = \square$

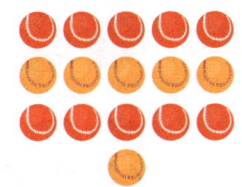
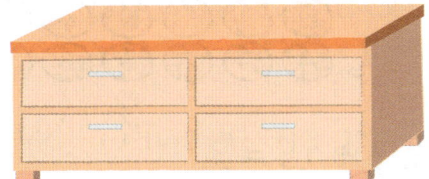

d 18 books
3 shelves
$18 \div 3 = \square$

➡ *Workbook page 57 and page 58*

Use arrays

Remember, an array is an arrangement in rows and columns.
We can use arrays to help us multiply.
We can also use them to help us divide.

Each array shows us a **fact family** of multiplication and division facts.

$3 \times 5 = 15$
$15 \div 5 = 3$

$5 \times 3 = 15$
$15 \div 3 = 5$

1 Say the fact family you can make with each array.

a

b

c

d

e

f
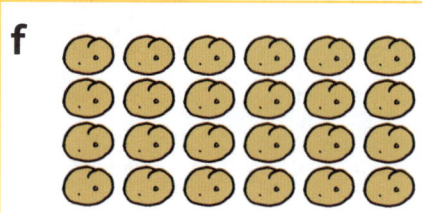

2 Chinua says that $15 \div 3 = 5$, so $3 \div 15 = 5$.
What is his mistake?

Divide and multiply

You can divide and multiply numbers in many ways:

Array

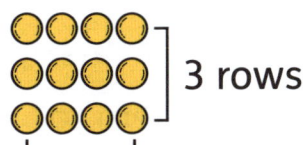

3 rows

4 in each row

$3 \times 4 = 12$
$12 \div 4 = 3$

Equal groups

3 groups
4 in each

$3 \times 4 = 12$
$12 \div 4 = 3$

Repeated addition

$4 + 4 + 4 = 12$
$3 \times 4 = 12$

Repeated subtraction

$12 - 4 = 8$
$8 - 4 = 4$ $12 \div 4 = 3$
$4 - 4 = 0$
Take away 4 three times.

Number line

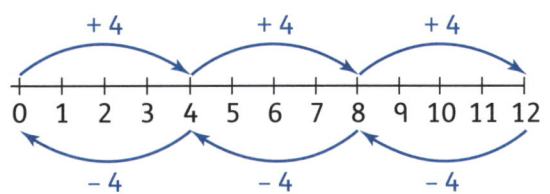

1 jump of 4 = 4 $3 \times 4 = 12$
2 jumps of 4 = 8 $12 \div 4 = 3$
3 jumps of 4 = 12

- Jump forwards to multiply.
- Jump backwards to divide.

1 How many times can you take away:

  a 5 from 20

  b 6 from 18

  c 4 from 24?

2 Use each answer from question 1 to help you answer these division questions.

  a $20 \div 5 = \boxed{}$ **b** $18 \div 6 = \boxed{}$ **c** $24 \div 4 = \boxed{}$

Solve division problems

1 Use the number line to help you.

```
0 1 2 3 4 5 6 7 8 9 10 11 12 13 14 15 16 17 18 19 20 21 22 23 24 25
```

a $15 \div 3 = \boxed{}$ b $16 \div 4 = \boxed{}$ c $20 \div 4 = \boxed{}$ d $21 \div 7 = \boxed{}$

2 Look at the numbers 2, 5 and 10. We can use them to make different division and multiplication facts. This is a fact family:

$2 \times 5 = 10$ $5 \times 2 = 10$ $10 \div 5 = 2$ $10 \div 2 = 5$

a Buhle says: 'This means that we can say: $2 \times 10 = 5$ and $5 \div 10 = 2$.' What are Buhle's mistakes?

b Choose a division or multiplication fact and use the numbers to write a fact family.

Problem solving

3 One flower has 5 petals. How many petals on:

a 3 flowers   b 7 flowers   c 5 flowers?

4 A flower seller has a bucket of 24 flowers. How many bunches can she make with:

a 6 flowers in each bunch   b 8 flowers in each bunch?

5 The flower seller sells pairs of sunflowers. A pair is 2. How many sunflowers does she need to make:

a 2 pairs   b 5 pairs   c 10 pairs?

First decide what you need to find out. Then choose a way to work it out.

➡ *Workbook page 59*

UNIT 13 Fractions

Whole and parts

> **Think and share**
>
> In the first bracelet, there are 3 beads.
> 1 out of 3 is black.
>
> Describe the other bracelets in the same way.
>
>
>
> For the first bracelet, we can write that $\frac{1}{3}$ of the beads are black.
> What could we write for the other bracelets?
>
> When you divide an object or group into equal parts, each part is called a **fraction**.

1 Each group has been split into 4 parts.
Say whether the parts are equal or unequal.

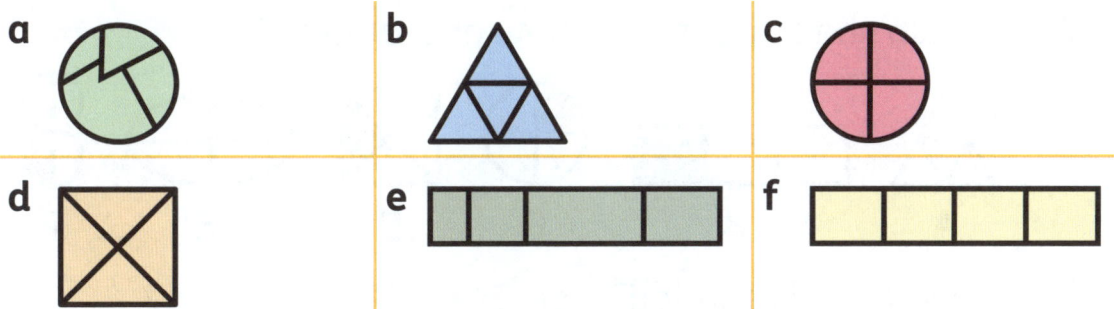

2 Each group has been split into 4 parts.
Say whether the parts are equal or unequal.

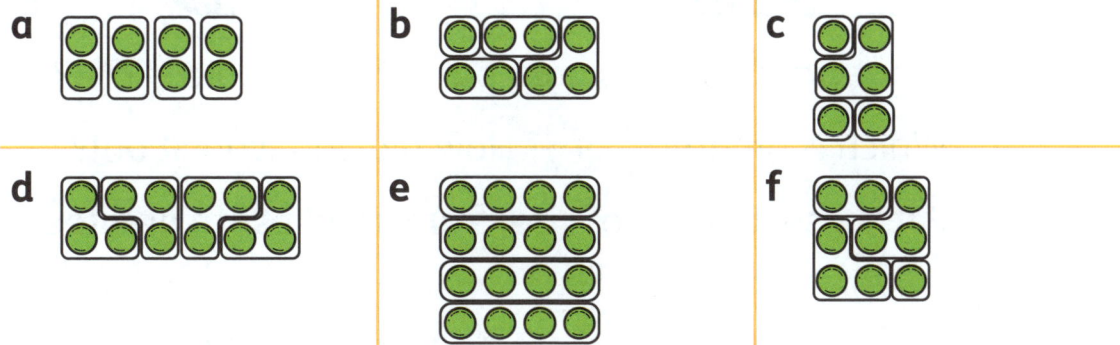

Halves and quarters

Each **whole** shape is divided into equal parts.

Each part is one-**half**.
1 out of 2 equal parts
$\frac{1}{2}$ = one-half
2 halves make 1 whole.
$\frac{2}{2}$ = 1

Each part is one-**quarter**.
1 out of 4 equal parts
$\frac{1}{4}$ = one-quarter
Four quarters make 1 whole.
$\frac{4}{4}$ = 1

1 Which shape does not show a half?

  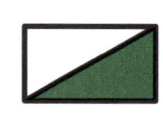  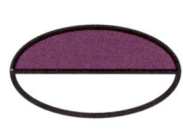 

A   B   C   D   E   F

2 Which shape does not show one-quarter?

    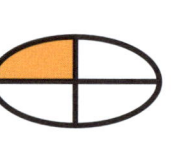

A   B   C   D   E   F

3 How much is shaded? $\frac{1}{2}$ or $\frac{1}{4}$?

a 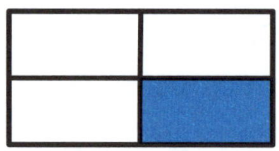   b 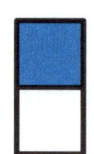   c    d

4 a What fraction is a bigger part of the whole: $\frac{1}{4}$ or $\frac{1}{2}$?

b Which is greater: $\frac{1}{2}$ or $\frac{3}{4}$? How did you work it out?

c Put these in order of size, from smallest to greatest:
$\frac{3}{4}$, 1, $\frac{1}{2}$, $\frac{1}{4}$

➡ *Workbook page 60 and page 61*

Division and fractions

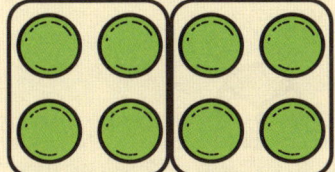  

$8 \div 2 = 4$ $\frac{1}{2}$ of $8 = 4$ Whole group = 8
1 out of 2 equal
parts = 4

1 What do you notice about division and fractions?
Tell your partner.

2 Work in pairs. Write the division fact and the fraction.
The first one has been done for you.

a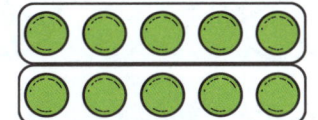

$10 \div 2 = 5$
$\frac{1}{2}$ of $10 = 5$

b

c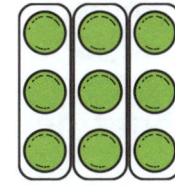

d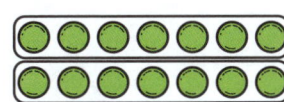

e

f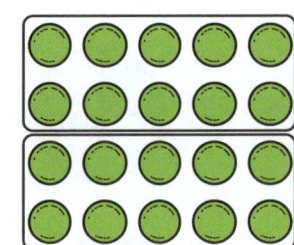

3 A bar of chocolate has 12 pieces. How many pieces does
each person get if you share the bar equally between:

a 2 people   **b** 3 people   **c** 4 people?

➡ *Workbook pages 62 to 64*

More about fractions

Each of these circles is divided into **thirds**.

one-third
$\frac{1}{3}$

two-thirds
$\frac{2}{3}$

three-thirds or one whole
$\frac{3}{3}$ or 1 whole circle

1 Which bar does not have one-third shaded?

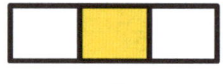   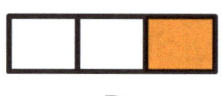      

A   B   C   D

2 Which circle does not have two-thirds shaded?

A    B    C    D

3 What fraction of each shape is shaded?
Choose the fractions from the box.

$\frac{1}{2}$ $\frac{2}{2}$ $\frac{1}{3}$ $\frac{2}{3}$ $\frac{1}{4}$ $\frac{2}{4}$ $\frac{3}{4}$ $\frac{4}{4}$ $\frac{3}{3}$

a    b    c

d    e 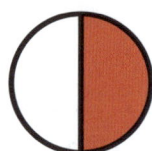   f

g    h    i

4 Which of the shapes from question 3 show the whole
shape shaded?

Equivalent fractions

Mark and Nala shared a pizza. They each got $\frac{1}{2}$.

Jan and Hannah shared a pizza. They each got $\frac{2}{4}$.

Can you see that 2 out of 4 equal parts is the same as one-half? Explain to a partner why this works.

$\frac{2}{4}$ and $\frac{1}{2}$ are called **equivalent fractions**.

1 Which of these fractions is not equivalent to the others? Why not?

A  B  C  D  E

2 Tell your partner about the different fractions. Find bars that show:

a halves

b quarters

c thirds.

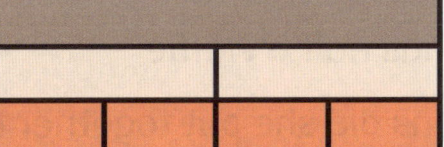

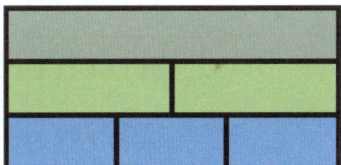

Problem solving

3 Malia had 4 toy cars. She gave 2 to her sister. Is each statement true or false?

a Malia gave half the cars to her sister.

b Malia kept $\frac{2}{3}$ of the cars.

c Malia divided the group of cars into two equal parts.

d Malia's sister got $\frac{1}{4}$ of the cars.

Put fractions together

A chef cooks waffles.
One whole waffle makes 4 quarters.
She uses quarters and halves to make different amounts on each plate.

A  **B**  **C**

We can put fractions together to make new fractions.

1 Look at plates A, B and C.

   a Which plates have less than a whole waffle?

   b Which plate has the equivalent of one whole waffle?

   c Which plate has $\frac{3}{4}$ of a waffle?

     Which fractions did she put together to make this amount?

   d Which plate has half a waffle?

     Which fractions did she put together to make this amount?

2 Draw three different ways you can make a whole waffle out of halves and quarters.

3 Put each group of shaded fractions together and say how much they make altogether.

  a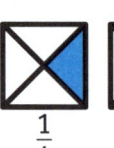

      $\frac{1}{4}$ $\frac{1}{4}$

  b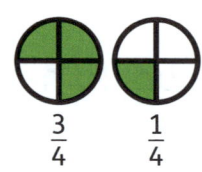

      $\frac{3}{4}$ $\frac{1}{4}$

  c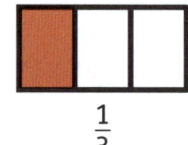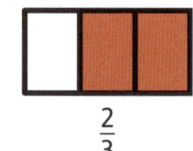

      $\frac{1}{3}$ $\frac{2}{3}$

➡ *Workbook page 65*

How long does it take?

Think and share

a  What can you do that takes less than 20 **seconds**?

b  What can you do for longer than 2 **minutes**?

1 How long does it take? Choose the correct time.

a  Counting to 60 slowly
... 58, 59, 60
1 second
1 minute

b  Turning the page
1 second
1 minute

c  Tying your shoelace
20 seconds
20 minutes

d  Washing your face
30 seconds
30 minutes

2 Tell a partner three things you do in the morning before you come to school. Say how long they take.

Problem solving

 You can add numbers in any order.

3 Kai spends 5 minutes washing, 20 minutes getting dressed and 15 minutes having breakfast. Then he takes 5 minutes to brush his teeth and close his bag.
How long does he take altogether?

Compare times

It takes about ...

$\frac{1}{4}$ of an **hour** to wash and dress.

$\frac{1}{2}$ an hour to eat dinner.

$1\frac{1}{2}$ hours to play a football match.

1 How long will it take? Choose the best estimate.

Remember, < means less than and > means greater than.

a

read one page
< or > $\frac{1}{4}$ of an hour?

b

eat a snack
< or > $\frac{1}{4}$ of an hour?

c

play a sports match
< or > an hour?

d

cook a simple meal
< or > an hour?

e

sweep the floor
< or > half an hour?

f

wait for the bus
< or > half an hour?

Seconds, minutes and hours

When you hear a clock ticking, each tick is one second.

60 seconds = 1 minute   30 seconds = $\frac{1}{2}$ a minute

60 minutes = 1 hour   30 minutes = $\frac{1}{2}$ an hour

1 About how long does it take? Choose the best estimate from the box.

1 second	3 seconds
$\frac{1}{2}$ minute	1 minute
1 hour	3 hours
$\frac{1}{2}$ hour	

a

I ride my bike.

b

I slide down the slide.

c

I write the number 1.

d

I play at the beach.

e

I wave goodbye.

f

I take a nap.

2 How many hours do you spend:

a sleeping at night   **b** at school during the day?

3 Write some more things you can do in:

a 1 minute   **b** 1 hour.

➡ *Workbook page 66*

Days and weeks

There are 24 hours in 1 **day**.
There are 7 days in 1 **week**.

In some places, people start the week on a Sunday.
In other places, the week starts on a Monday.

1 These are the days of the week. They are all mixed up.
Can you say them in the correct order?

| Thursday | Sunday | Monday | Friday |

| Tuesday | Saturday | Wednesday |

2 Say the name of:

a the day before Monday

b the day after Thursday

c the day before Wednesday

d the day after Friday.

3 The pupils in this Year 2 class voted for their favourite day of the week.

a How many pupils like Wednesday best?

b Which day had half as many votes as Wednesday?

c Which two days had equal votes?

d Which was the most popular day?

Favourite days of the week	
Monday	☺
Tuesday	☺ ☺ ☺
Wednesday	☺ ☺
Thursday	☺ ☺ ☺
Friday	☺ ☺ ☺ ☺ ☺ ☺ ☺
Saturday	☺ ☺ ☺ ☺ ☺ ☺
Sunday	☺ ☺ ☺ ☺

Key: ☺ = 1 pupil

Months and years

One **year** is 365 days or 12 **months**. Each month has about 30 days. That is about 4 weeks. February has 28 days, but every fourth year is a **leap year**. In a leap year, February has 29 days.

January 31 days	February 28 or 29 days	March 31 days	April 30 days	May 31 days	June 30 days
July 31 days	August 31 days	September 30 days	October 31 days	November 30 days	December 31 days

1 Say the names of the months in order.

2 Look at the numbers of days in the months. Is there a pattern? Say what you notice.

3 Here is a way to remember the days of the months:

- Make knuckles. Can you see mountains (the knuckles) and valleys (the dips between)?

- Say the month names as you point on top of each 'mountain' or 'valley'.

- The mountains have 31 days.

- The valleys have 30, except for February, which has 28 or 29.

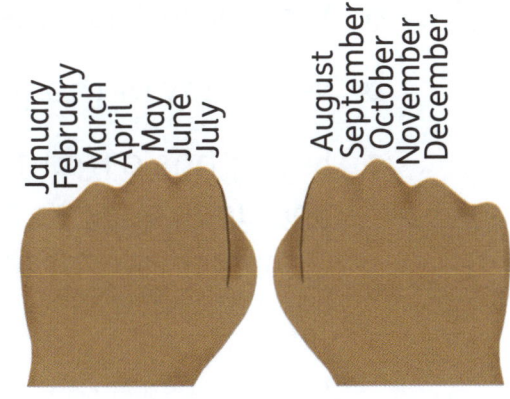

 Problem solving

Every fourth year is a leap year.

4 The years 2016 and 2020 were leap years. Are we in a leap year this year? How can you find out?

➡ *Workbook page 67*

Use a calendar

A **calendar** is a chart that helps us to remember **dates**.
A date is a specific day in a month.

OCTOBER						
M	Tu	W	Th	F	Sa	Su
		1	2	3	4	5
6	7	8	9	10	11	12
13	14	15	16	17	18	19
20	21	22	23	24	25	26
27	28	29	30	31		

1 On which day of the week is:

  a 1st October

  b 14th October

  c 31st October?

2 In the month shown, which dates fall on a:

  a Monday   **b** Sunday?

3 What day of the week is it:

  a a week after 7th October

  b 2 weeks before 25th October?

4 **a** What was the date on the day before 1st October?

  b What day of the week is 1st November?

  c Skylar has a mid-term break that starts on 3rd October.
    She goes back to school 10 days later.
    On which day of the week does she go back to school?

> Think about the number of days in the months and the weeks.

Problem solving

5 Is it ever possible that there are only 3 Tuesdays in a month?
  How can you work it out?

Compare and sequence

60 seconds = 1 minute
60 minutes = 1 hour
24 hours = 1 day

7 days = 1 week
about 4 weeks = 1 month
12 months = 365 days = 1 year

1 How many days is:

    a 2 weeks   **b** 3 weeks   **c** the month of January?

2 How many weeks is:

    a 2 months   **b** the month of February?

3 These children are holding up their times for completing a race. Write their names in order from the fastest to slowest.

> Think about how many seconds are in a minute.

Nirman
1 minute and 10 seconds

Abdul
1 minute and 5 seconds

Rosie
55 seconds

Mina
61 seconds

Sam
1 minute

💡 Problem solving

4 These children have signs that show how long it is until their eighth birthdays.
Write their names in order from youngest to oldest.

> The youngest has the longest time to wait until their birthday.

Callum
3 weeks

Joe
2 years

Angela
8 months

Dev
one and a half years

Hanif
4 weeks

➡ *Workbook page 68*

UNIT 15 Possible outcomes

What is the possible outcome?

💭 Think and share

Which colour is it **possible** to pick from the jar?
Which colour is it **impossible** to pick?

blue bead   white bead   black bead

An **outcome** is something that could happen.
You could pick a blue bead or a black bead from this jar.
Those are the possible outcomes. Something is possible if it can happen. It is impossible if it cannot happen.

1 Are these events possible or impossible?

I will see a flying elephant today.

There will be clouds tonight.

I will visit the moon next Tuesday.

I will read a book this week.

2 Look at the spinner. Are these outcomes possible or impossible? Explain how you know.
- The pointer lands on green.
- The pointer lands on red.
- The pointer lands on yellow.

3 A drawer has only plain black and plain white socks. Are these outcomes possible or impossible?
- You pull out one black sock.
- You pull out one striped sock.
- You pull out one white sock.

➡ *Workbook page 69*

Regular patterns

Some events follow a regular pattern. This means they follow a rule. It repeats in a way that makes it easy for us to say what will come next.

For example, day follows night and night follows day.

day   night   day   night   day

Number patterns can also repeat. For example:

6   7   6   7   6   7

They can also follow a rule, such as adding or subtracting. For example:

3   6   9   12   15

Start at 3 and add 3 to get to the next number.

1   Does it follow a regular pattern? Discuss with a friend.

   a The days of the week   **b** The weather

   c The hours of the day   **d** The five times table.

2   These patterns are regular. Describe each pattern to a partner. Say the rule and what should come next in the pattern.

   a

   b | | | | | | | | | | |

   c 10   20   30   40   50   60

   d 99   97   95   93   91

   e

Random patterns

Some events are **random**. Think about watching cars drive past. If you write down the different makes of cars, you won't find a pattern in the makes. You can't work out what make the next car will be.

Look at this number pattern:

3   5   1   2   4   5   1   9   8   5   1

Sometimes we see some rules in a random pattern.
For example, in this pattern a 1 always comes after a 5.

1 These patterns are random.
Describe them.
Say whether you notice any rules.

a

b

c   A B R G B R F Z L T B R

d   2 1 1 2 1 1 1 2 2 2 1 2 1 2 1 1 2 1 2 1 2 2 2 1 2 1 2 2 1 1 1 2 1 2 2

e

💡 **Problem solving**

2 Use four squares and four circles.
Make two different regular patterns
and two different random patterns.

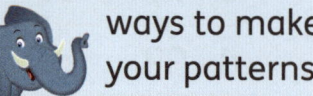
You can cut out the shapes and arrange them in different ways to make your patterns.

➡ *Workbook page 70*

Experiments

Ellie asks: 'If I flip a coin 20 times, how many times will it land on heads, and how many times on tails?'

An **experiment** is a kind of test that you do, to see what will happen.

To record means to write down.

The **results** are what happens in the experiment.

heads   tails

Heads	Tails
卌 卌 II	卌 III

| = 1

卌 = 5

1   a   What are the possible outcomes of flipping a coin?

    b   How many times did Ellie's coin land on heads?

    c   How many times did it land on tails?

2   a   What do you think Ellie did in her experiment?

    b   If you flipped a coin 20 times, do you think you would get the same results as Ellie? Why or why not?

3   Do your own coin experiment:

- Check which side is heads and which is tails.

- Flip the coin 20 times and see how it lands.

- Record your results in a table.

4   Make a spinner with two colours like this. Spin the spinner 20 times and record your results in a table.

green

yellow

Mixed practice 2

1   At a hotel, there are five kinds of fruit for breakfast.

Fruits eaten at breakfast

a   Which fruit did most people choose?

b   How many people chose it?

c   How many more people chose banana than orange?

d   The hotel only wants to serve three fruits.
Which should they choose? Why?

2   **Sports choices**

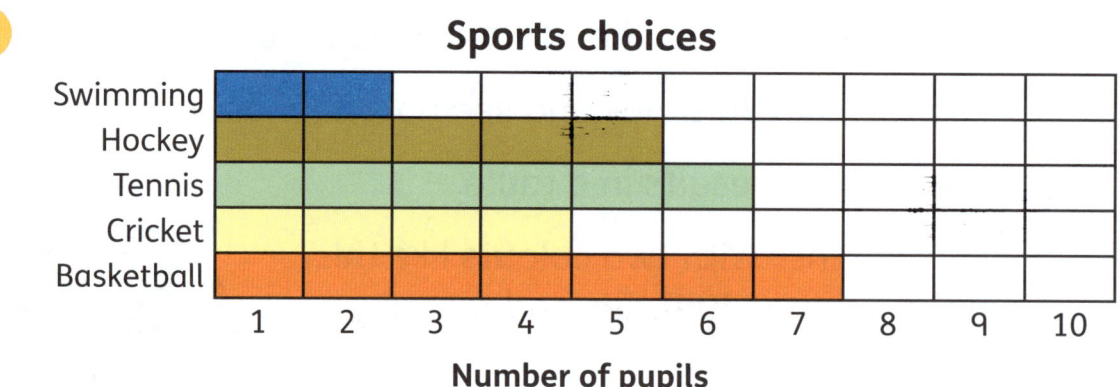

Make up three questions that you can ask about this
block diagram.

3 Write the fact family of ÷ and × facts for each array.

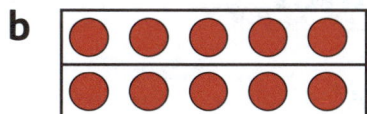 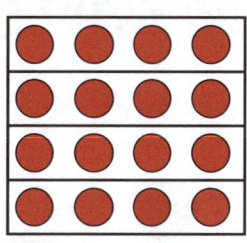

a   b   c

4 Are these calculations correct or incorrect? Explain.

    a $7 \times 2 = 2 \times 7$ **b** $\frac{1}{2}$ of $8 = 8 \div 2$ **c** $6 \div 3 = 3 \div 6$

5 **a** There are 20 pupils in a class. The teacher asks them to sit in groups of 4. How many groups do they make?

    b Some more pupils joined the class. There are now 8 groups of 4. How many more pupils joined?

6 Draw your own shapes and shade them to show:

    a $\frac{1}{4}$ **b** $\frac{1}{2}$ **c** $\frac{2}{3}$

7 Complete the sentences.

There are ☐ days in a week and ☐ weeks in a month.

There are ☐ minutes in an hour and ☐ minutes in a half hour.

8 Name an activity you can do in:

    a 1 second   **b** 5 minutes   **c** 1 hour.

9 What are the possible outcomes of flipping a coin?

10 Complete each sentence.

    a Today it is possible that I will …

    b Today it is impossible that …

Symmetry

Investigate symmetry

 Think and share

A

B

How were these shapes made? What do you notice about them?

We call the fold line in these pictures a **mirror line**.
Why do you think we use that name?

1 Make your own mirror picture. You will need: paper, paint and paintbrushes.

- Fold a sheet of paper in half.
- Paint some blobs of colour on one half.
- Fold it over to print.
- Open it out to see what you have made.

Try this with different designs.

2 Make a new picture. This time, you will need: paper, markers and a pair of scissors.

- Fold a sheet of paper in half.
  Draw a shape along the fold line.
- Cut along the lines you have drawn.
  Do not cut along the fold.
- Unfold the shape.
  What have you made?

fold

cut out

draw

Line symmetry

These shapes have **line symmetry**. This means that you can draw a line through them to divide them into two parts that match exactly.

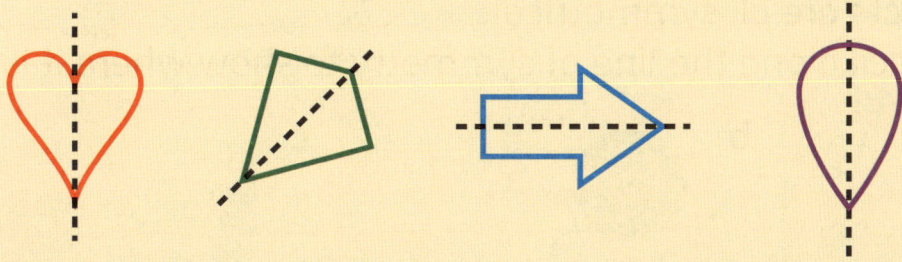

The dotted line is called a mirror line or a **line of symmetry**. When you fold a shape along this line, both sides fit onto each other exactly.

1 Bella tried to draw lines of symmetry, but she made some mistakes. Which lines are not lines of symmetry?
Where is the line of symmetry on each shape?
Show by placing a pencil in the correct place.

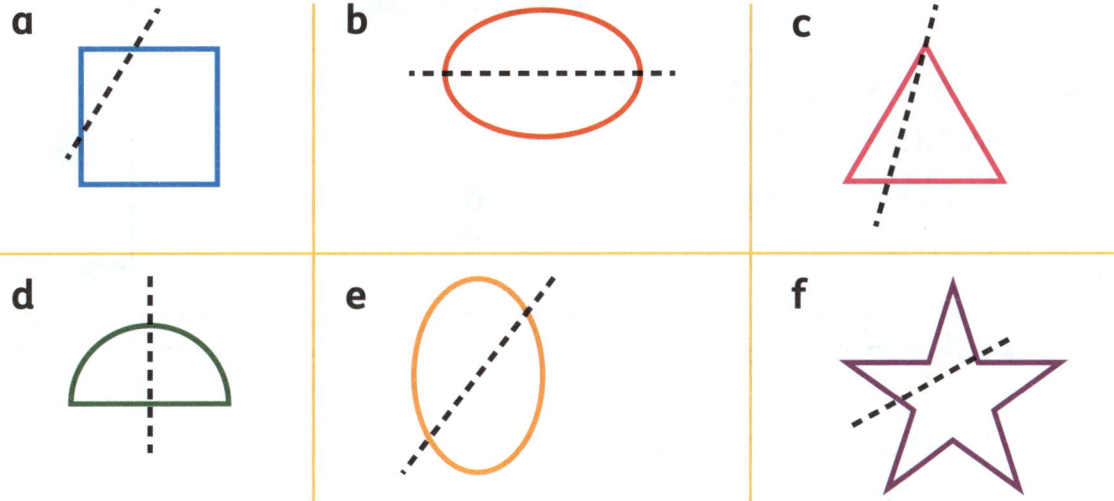

2 Work with a partner.
Draw three different **symmetrical** shapes.
How can you prove that your shapes are symmetrical?

➡ *Workbook page 71*

Symmetrical objects

When a shape or object has line symmetry, it is symmetrical.

1 These objects are all symmetrical.
Place a pencil along the line of symmetry to show where it is.

2 These shapes are not
symmetrical.
Explain why not.

 Problem solving

3 Leah likes to arrange her eggs
symmetrically.

   a Draw some other ways she could
arrange 8 eggs symmetrically in
this tray.

   b Try working out symmetrical arrangements for other
numbers of eggs, for example, 12 eggs.

Reflections

Look what happens when you place a mirror upright along a line of symmetry.
Each half of the shape is a **reflection** of the other.

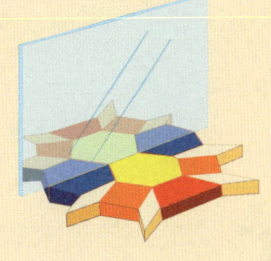 

1 Create your own mirror pictures.
You will need: a small mirror and some pattern blocks or other flat shapes.
- Build a shape with one side against the mirror.
- Use the mirror to see the whole symmetrical shape.
- Draw a picture of the symmetrical shape you have made.

2 Copy these shapes.
Complete the reflection of each shape.
Say the letter or

 Problem solving

 Use capital letters.

3 Find some more numbers or letters that have a line of symmetry.
Draw a mirror line and complete one side of the number or letter.
Let your partner complete the other side.

➡ *Workbook page 72 and page 73*

Capacity and temperature

Compare containers

Think and share

Gaia found some water containers in her house.

water refill bottle	Lexi's	Dina's	Gaia's	Gaia's baby sister's

Which container can hold the most water?
Which can hold the least water?

1 Compare the drink bottles of pupils in your class.

    a Estimate how much each bottle holds.

    b How could you measure how much each bottle holds?

2 Work in pairs.
Estimate how many cups
of juice you can pour
from each of these
containers.
Share your ideas
with the class.

Units of capacity

Capacity is the amount of liquid that a container can hold.

We use **millilitres** (ml) to measure the capacity of small containers.

We use **litres** to measure the capacity of larger containers.

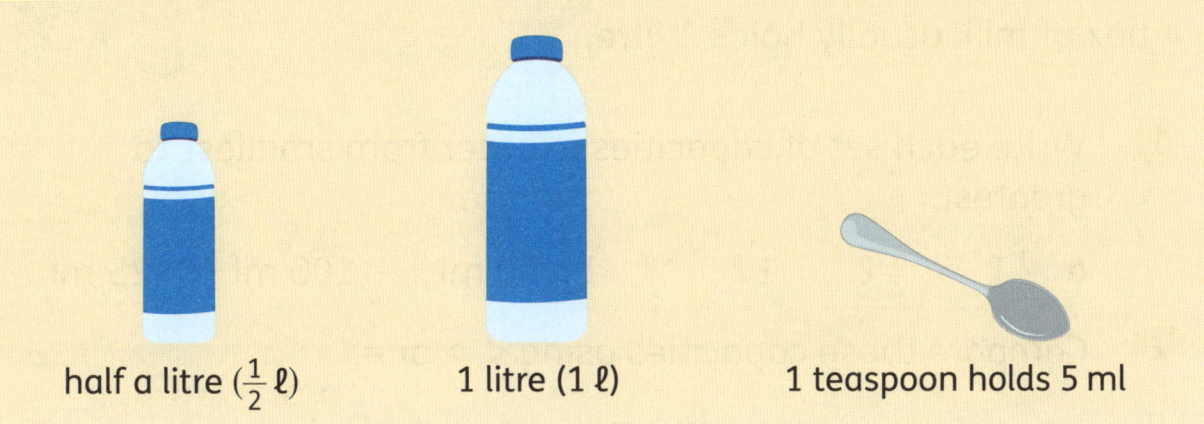

half a litre ($\frac{1}{2}$ℓ)   1 litre (1 ℓ)   1 teaspoon holds 5 ml

1   Would you use litres or millilitres to measure the capacity?
Give reasons.

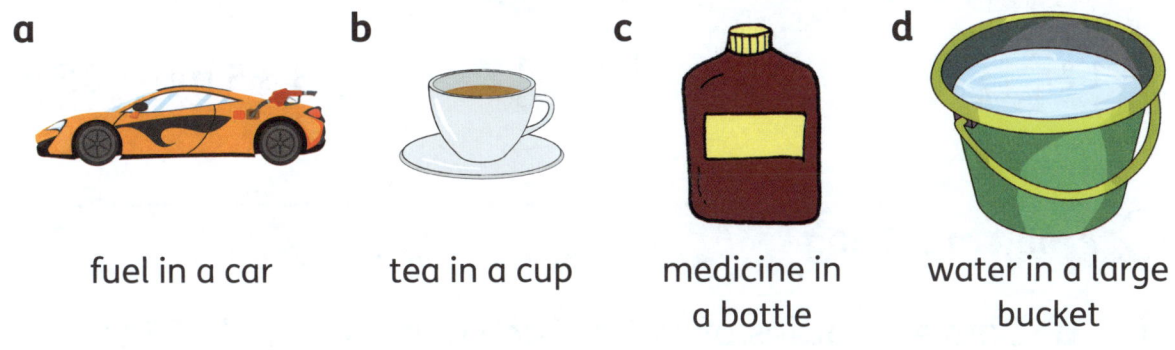

a   fuel in a car   b   tea in a cup   c   medicine in a bottle   d   water in a large bucket

2   Choose the best estimate for the capacity of each container.

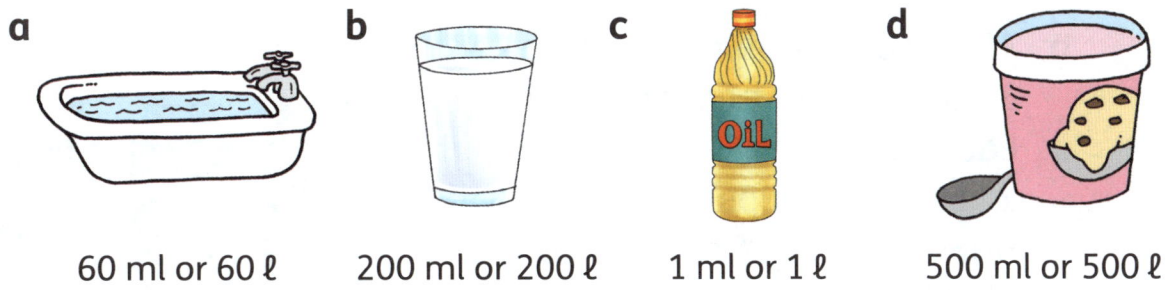

a   60 ml or 60 ℓ   b   200 ml or 200 ℓ   c   1 ml or 1 ℓ   d   500 ml or 500 ℓ

➡ *Workbook page 74*

Working with capacity

Remember, we use millilitres for small amounts.

It takes about 20 drops of water to make 1 ml.

We use litres for bigger amounts.

A box of milk usually holds 1 litre.

1 Write each set of capacities in order from smallest to greatest.

a $\frac{1}{2}$ℓ $\frac{1}{4}$ℓ   1ℓ   b 50 ml   100 ml   25 ml

2 Compare these capacities using <, > or =.

a

b

c

1 ℓ ☐ $\frac{1}{4}$ ℓ $\frac{1}{2}$ ℓ ☐ $\frac{1}{3}$ ℓ   3 × 5 ml ☐ 15 ml

---

💡 Problem solving

3 **a** I have 3 bottles of water. Each bottle can hold 1 litre.
One bottle is full, and the other two are half full.
How much water do I have?

b At a sports match, the coach brings
3 full boxes of juice for the teams.
Each box holds 1 litre.
1 litre fills 4 cups.
How many cups can the coach
fill with juice?

➡ *Workbook page 75*

Temperature

Think and share

Ava is ill.
What is in her mouth?
What does it do?
What happens to your **temperature** when you are ill?

The air inside a freezer is cold.
The steam from a kettle is hot.
When we measure how hot or cold things are, we measure their temperature.

1 Which are hot, and which are cold?

a 	**b** 	**c** 	**d** 
e 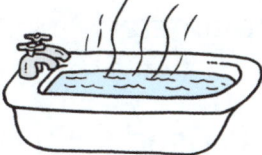	**f** 	**g** 	**h** 

2 In each pair, which is:

a hotter

b colder?

➡ *Workbook page 76*

Measure temperature

We use a **thermometer** to measure temperature.
It shows the temperature in units called **degrees Celsius** (°C).

Glass thermometers have a special liquid inside.
As the temperature goes up, the liquid expands and moves up the tube.

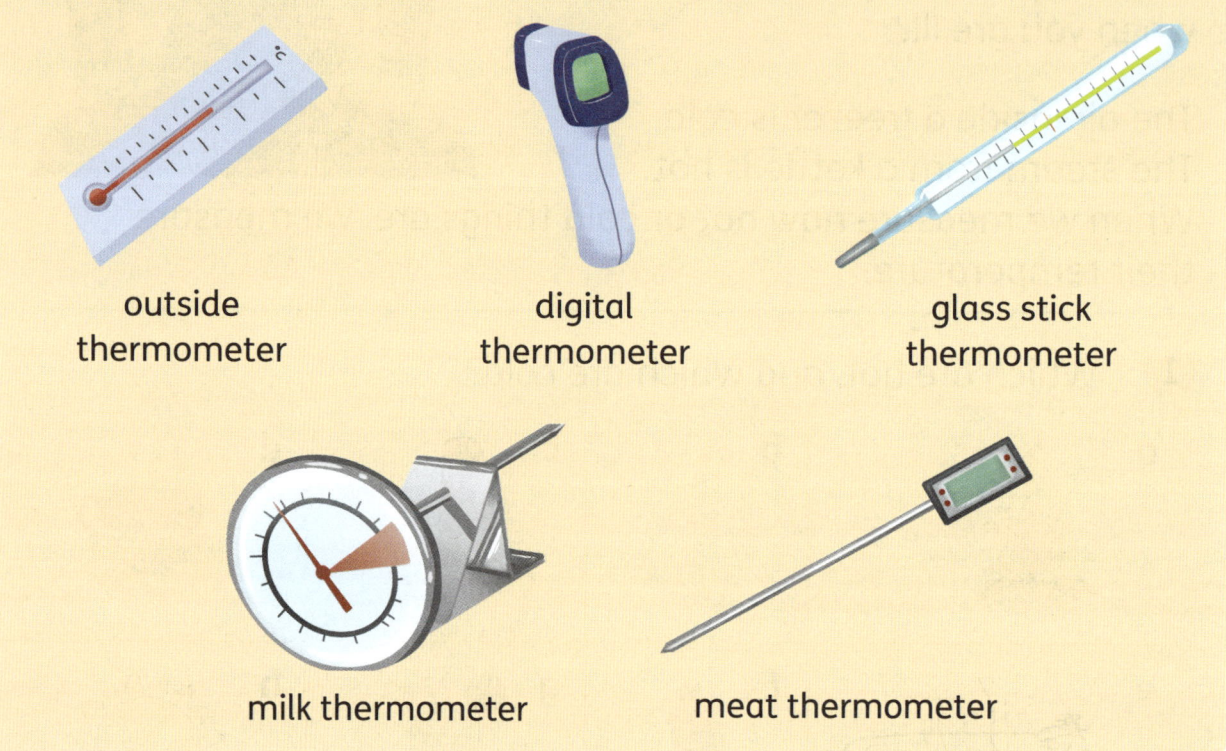

outside
thermometer

digital
thermometer

glass stick
thermometer

milk thermometer

meat thermometer

1 **a** What do you think each thermometer is used for?

b Which kinds have you seen at home or at school?

c Why do you think we use digital forehead thermometers more than glass stick thermometers these days?

d Look at the scale on a stick thermometer.
How is it similar to a ruler?

2 Your teacher will bring a thermometer to class.
Estimate and then measure some temperatures of people and things in your class.

Time on the clock

💭 **Think and share**

What numbers are usually on the face of a clock? What is different about clock B? Why are there extra numbers?

A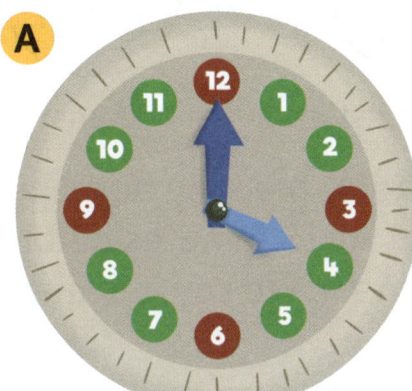

B

How could clock B help you learn to tell the time?

There are 60 minutes in an hour.
On a clock, the short hand points to the hour.
The long hand shows the minutes past or before the hour.

3 o'clock

30 minutes past 7 is half past 7.

half past 7

1 Read the time on these clocks.

a    b 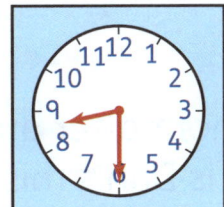   c    d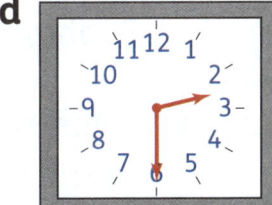

➡ *Workbook page 77*

Quarter hours

Remember, there are 60 minutes in an hour. The long hand starts each hour on the 12. One-quarter of an hour is 15 minutes.

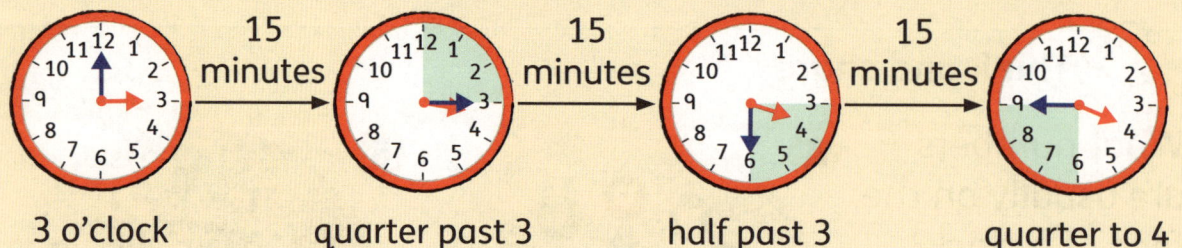

3 o'clock   quarter past 3   half past 3   quarter to 4

At 15 minutes past the hour, the long hand points to the 3. We say it is **quarter past**. At 15 minutes before the next hour, the long hand points to the 9. We say it is **quarter to** the hour.

1 Look at the clocks above.

    a What does the shaded part show?

    b How many quarter hours are there in one hour?

    c What will the time be 15 minutes after quarter to 4?

2 Read the times on these clocks.

a  **b**  **c**  **d**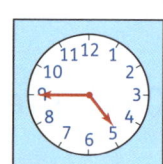

e  **f** 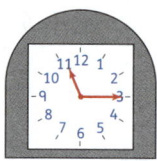 **g**  **h**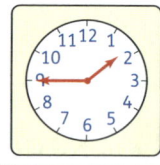

Draw a picture to help you.

💡 **Problem solving**

3 Dan says: 'Quarter of an hour after quarter past 6, the time is half past 9.' What is Dan's mistake?

➡ *Workbook page 78*

Digital clocks

Digital clocks show the time electronically, usually with numbers and dots.
This is how we often see the time on a digital watch, phone or computer.

8 o'clock   half past 12

The two dots in the middle are called a colon.
The number before the colon tells you the hour.
The number after the colon tells you the minutes.

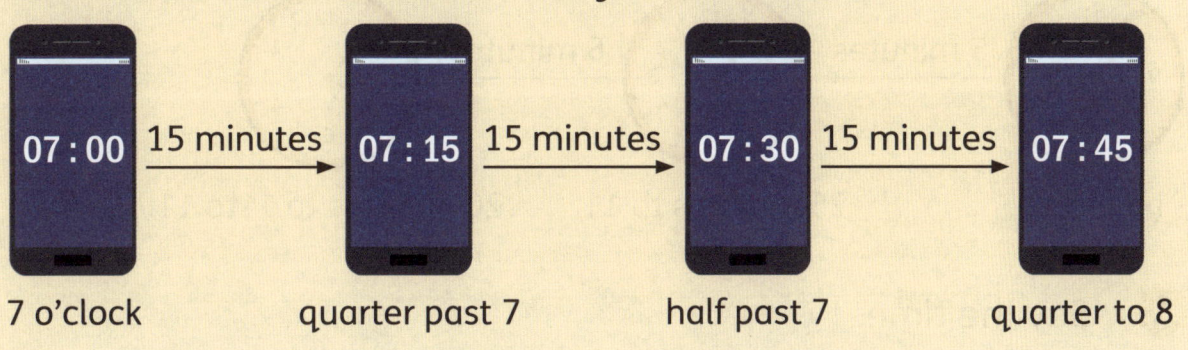

7 o'clock   quarter past 7   half past 7   quarter to 8

1 Read the time on each phone, watch and clock.

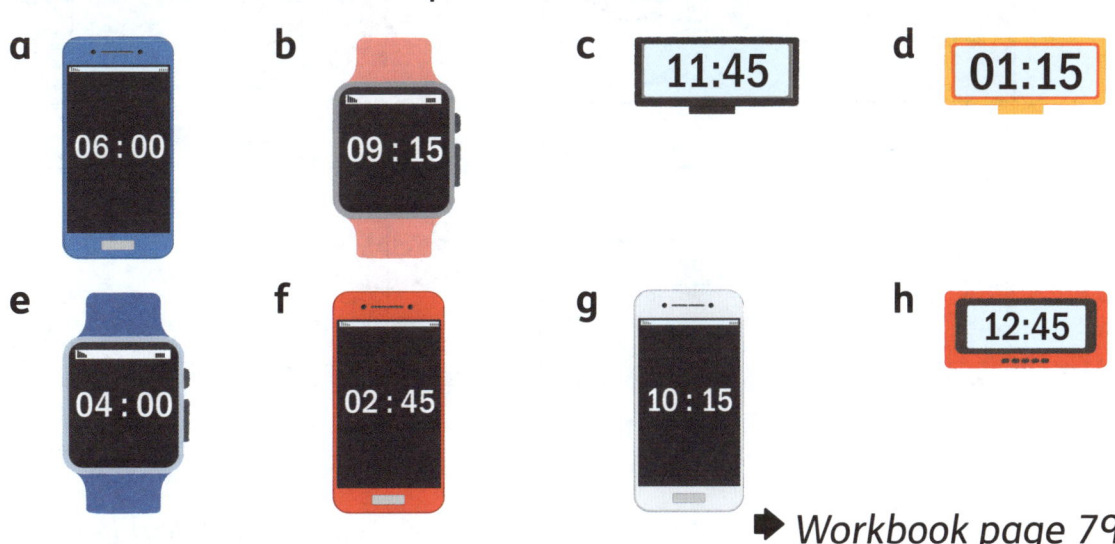

a 06:00   b 09:15   c 11:45   d 01:15

e 04:00   f 02:45   g 10:15   h 12:45

➡ *Workbook page 79*

More about clocks

It takes 60 minutes for the long hand to go all the way around the clock face.
After 5 minutes, the long hand points to the 1.
We say the time is 5 past the hour.
After another 5 minutes, we say the time is 10 past the hour.

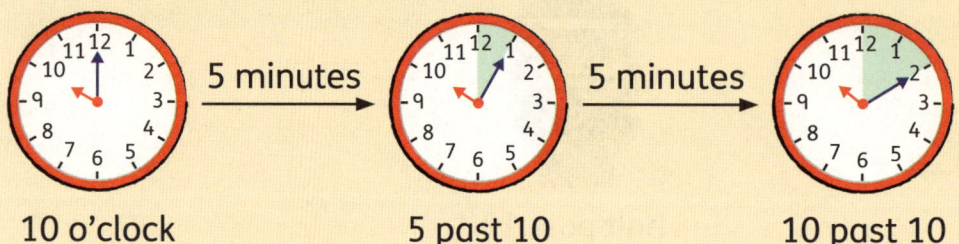

10 o'clock   5 past 10   10 past 10

After the clock gets to half past the hour, we say time differently.
We say how many minutes until the next hour.

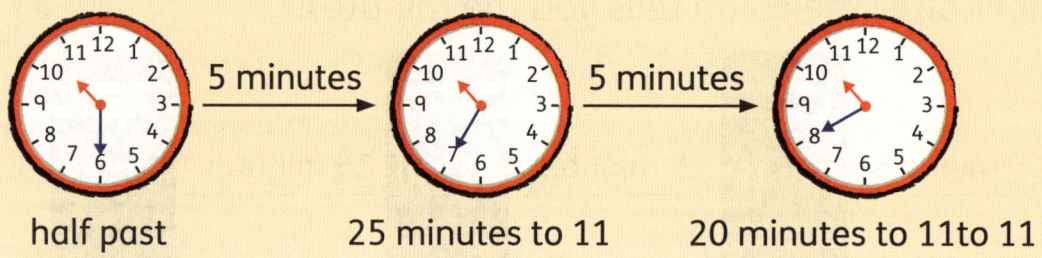

half past   25 minutes to 11   20 minutes to 11to 11

1   Say the time.

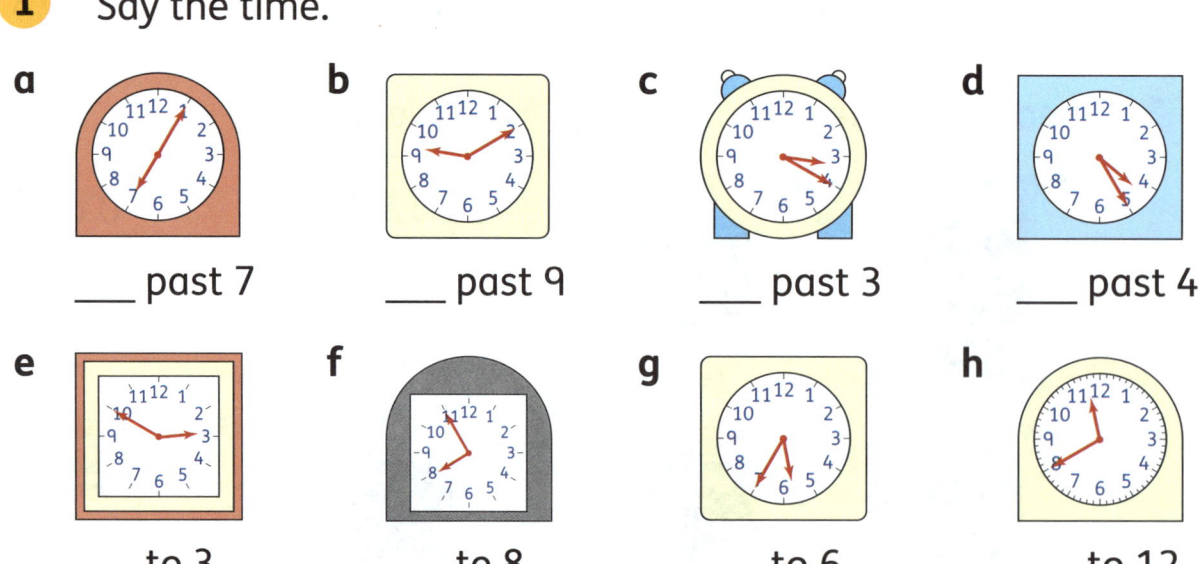

a ___ past 7   b ___ past 9   c ___ past 3   d ___ past 4

e ___ to 3   f ___ to 8   g ___ to 6   h ___ to 12

➡ *Workbook page 80*

More digital times

With digital times, the number after the colon tells us how many minutes have passed.
After the minutes pass 30, we need to work out how many minutes until the next hour.

20 minutes past 7

25 minutes to 8

1 Look at this picture.
How does it help you to work out how many minutes have passed when the minute hand is pointing at each number?

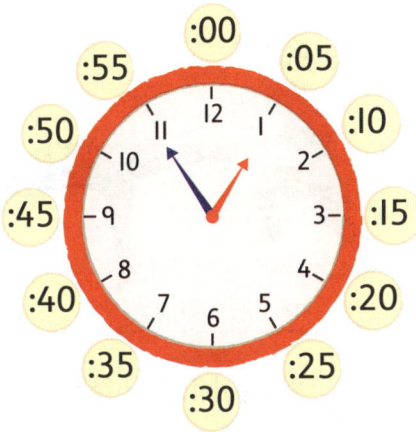

:00
:55 :05
:50 :10
:45 :15
:40 :20
:35 :25
:30

2 Match each digital clock to the clock face showing the same time.

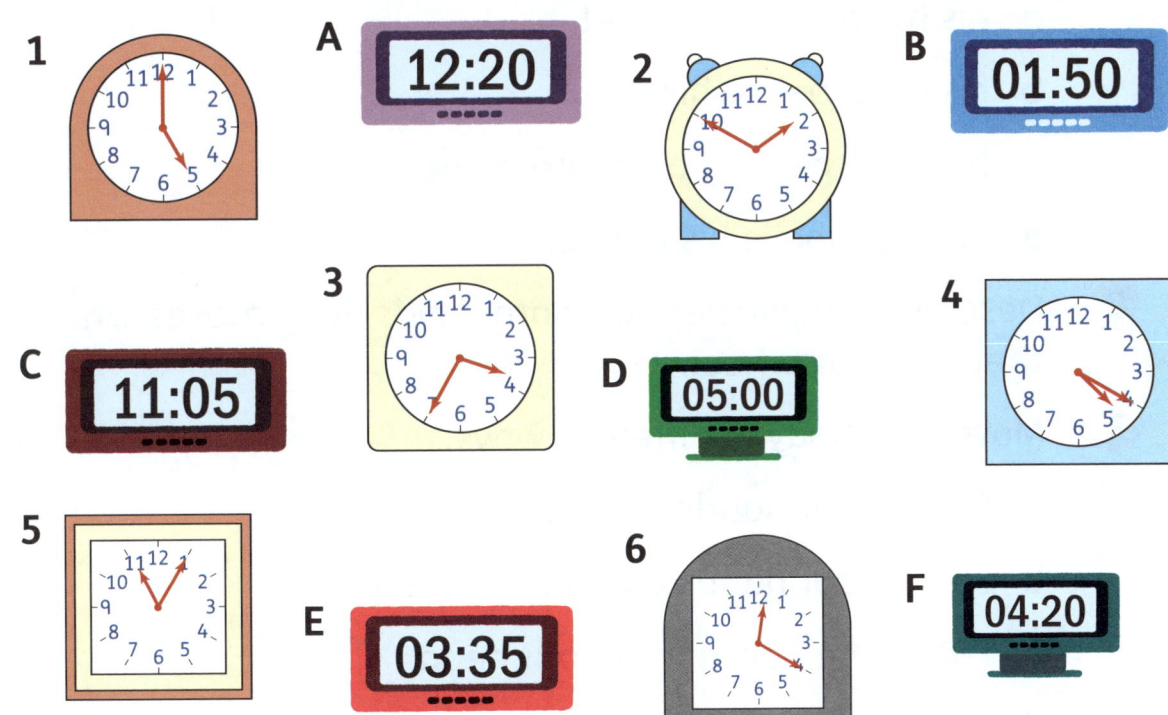

1

A **12:20**

2

B **01:50**

3

C **11:05**

D **05:00**

4

5

E **03:35**

6

F **04:20**

➡ *Workbook page 81*

Position and movement

Turning shapes

> ☁️ **Think and share**
>
> Can anyone in your class do a cartwheel?
> Talk about what this boy is doing with his body in each picture.
>
>

1 In which picture or pictures does he have:

   a his head above his feet

   b his head below his feet

   c his body making a square corner

   d his legs above his hands?

2 Describe to a partner how someone's body moves when they do a cartwheel.

3 Move your body in different ways.

   a Shake your hands up above your head.

   b Shake your hands down to your feet.

   c Put your elbow above your head.

   d Put your knee higher than your hip.

   e Put your shoulders lower than your feet.

Moving in a straight line

We can move an object in straight lines.
We can move **left** or **right, up** or **down**.
We can move **backwards** or **forwards**.

Use a counter. Start at the shoe.

1 Follow the directions and say which picture you get to.

  a Move 3 blocks up to get to the _____.

  b Turn right. Go forwards 3 blocks to get to the _____.

  c Continue for 2 blocks. Then go down 2 blocks to get

  to the _____.

2 Give your own directions to get:

  a from the boat to the treasure chest

  b from the tree to the bucket

  c from the bucket to the cup.

➡ *Workbook page 82*

Turns

The **direction** of a turn can be **clockwise** or **anti-clockwise**.

clockwise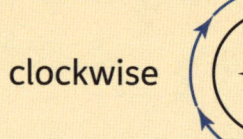

anti-clockwise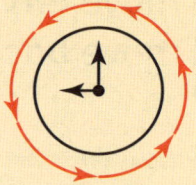

The hands of a clock turn clockwise.

Anti-clockwise is the opposite direction.

The amount of turn is how far round it goes.

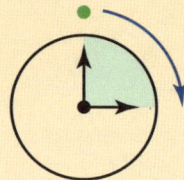

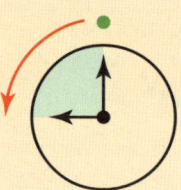

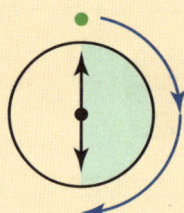

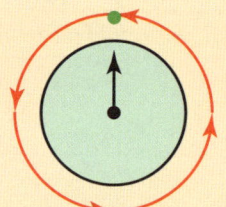

quarter turn clockwise

quarter turn anti-clockwise

half turn clockwise

full turn anti-clockwise

💡 Problem solving

1 Danielle made this pinwheel.
She always starts it in this position.
Which colour and pattern point up after:

a a quarter turn clockwise

b a quarter turn anti-clockwise?

2 Which turns could result in these end positions?

a
b
c

➡ *Workbook page 83*

More turns

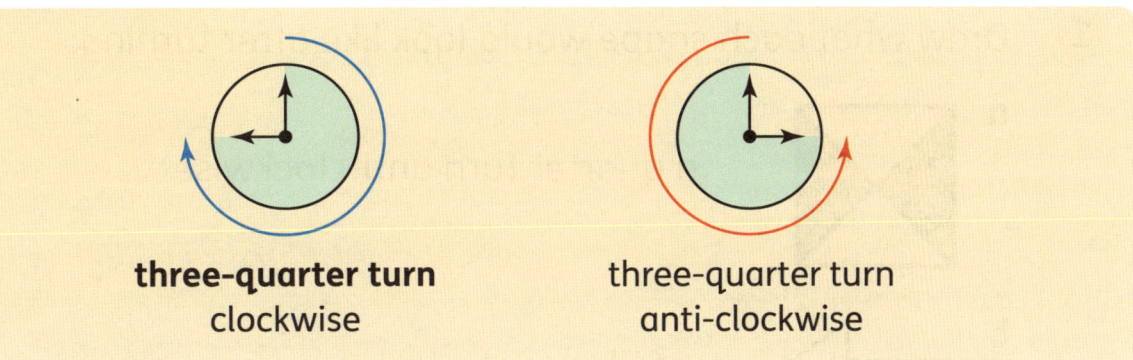

three-quarter turn
clockwise

three-quarter turn
anti-clockwise

1 How can you turn each shape to move it from the start position to the end position?

- First, choose the direction – clockwise or anti-clockwise.

- Then choose the size of the turn – quarter turn, half turn, three-quarter turn or full turn.

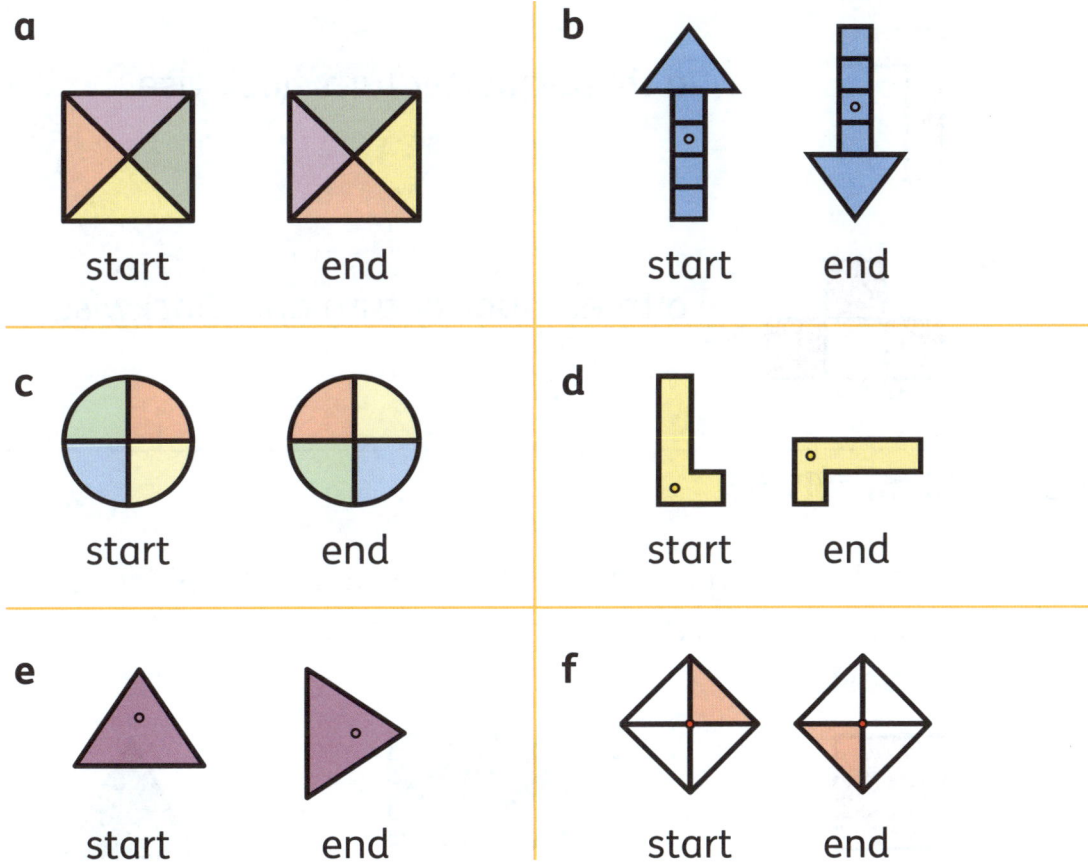

a

start   end

b

start   end

c

start   end

d

start   end

e

start   end

f

start   end

➡ *Workbook page 84 and page 85*

Working with turns

1 Draw what each shape would look like after turning:

a    a quarter turn anti-clockwise

b 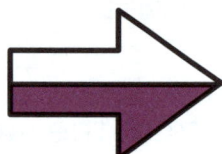   a full turn clockwise

c    a half turn clockwise

d    a three-quarter turn clockwise

e    a three-quarter turn anti-clockwise.

 Problem solving

2 Imagine turning each of these shapes one full turn. How many times during the turn would it look like its start position?

a    **b**    **c**

UNIT 20 Money

Coins and notes

Think and share

Do any of these notes and coins look like the money you use in your country?
What looks the same?
What looks different?

We use money to pay for things. Cash means money that we have ready to use, such as notes and coins.
Notes are made out of paper or plastic.
Coins are made out of metal.
The value of a coin or a note tells us how much it is worth.

1 Look at some of the money used in your country.
Draw pictures of:

a three coins that have a low value

b two coins that have the highest value

c three different notes.

2 Imagine that some pupils set up a snack stand at your school.
Draw a picture of some of the snacks and drinks they could sell. Suggest a price for each item.

SNACK STAND

Make different amounts

A **currency** is a system of money that people use in a particular place.
Some currencies used around the world include:

| British pound | European euro | Indian rupee | Malaysian ringgit | US dollar |

1 In the USA, people use dollars ($) and cents (c).
There are 100 cents in $1.
With a partner, work out which hands are holding $1.

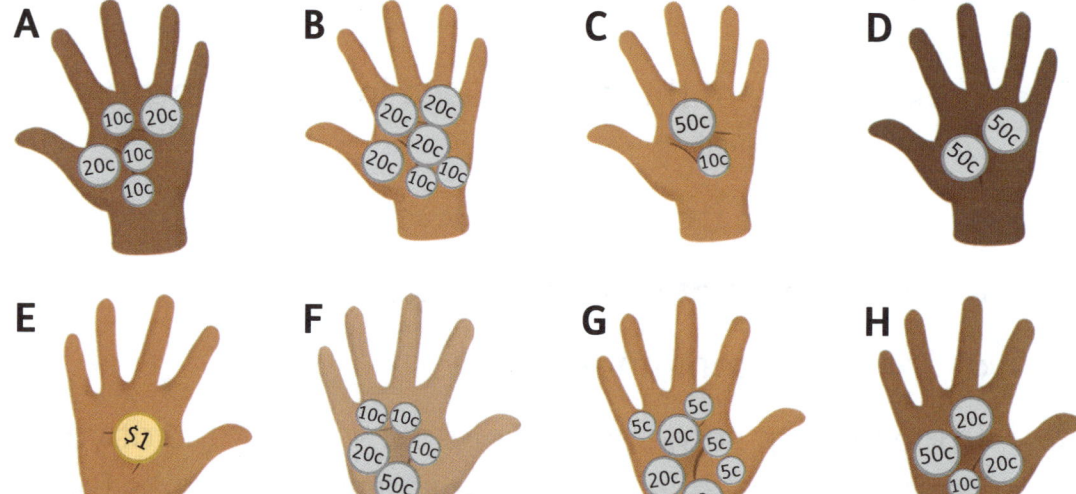

2 Make a poster about the currency in your country.

a Show the main coins and notes that you use.

b Choose one of the notes.

Draw different ways of making the same amount using coins.

➡ *Workbook page 86*

Pounds and pence

In the UK, people use the British pound.
This currency is also called pound sterling.

one pound (£1) = one hundred pence (100p)

Here are some British coins:

| 1p | 2p | 5p | 10p | 20p | 50p | £1 | £2 |

Here are some British notes:

1 Add up the coins or notes.
What coins or notes are missing?
The first one has been done for you.

a £2 + £2 + £1 = £5

b ? = 20p + 20p + 10p

c 50p + ? = £1

d £10 = ? + ?

e £20 = ? + ? + £5

The value of notes

The price is $5.
I pay with this note:

$20 is too much.
The shopkeeper gives me **change**.

$20 − $5 = $15
The change is:

1 What is the value of each set of notes?

a | 10 | 5 | b | 10 | 20 |

c | 50 | 20 | d | 20 | 20 | 5 |

💡 Problem solving

2 How much change should I get?

a I buy a T-shirt for $13.
I pay with:

| 10 | 5 |

b I buy some groceries for $24.
I pay with:

| 10 | 20 |

3 How much more do you need to make $100?

a | 10 | 20 | 20 |

b | 20 | 20 | 20 |

c | 20 | 50 | 20 |

➡ *Workbook page 87 and page 88*

Mixed practice 3

1 **a** Which of these shapes is not symmetrical?
Explain your answer.

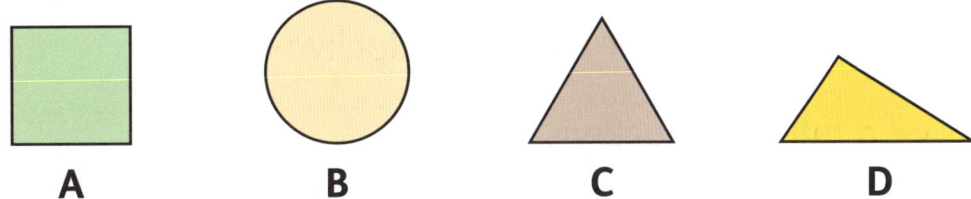

A   B   C   D

b Show the line of symmetry on each of the other shapes.

2 For each container, say whether it holds 1 litre, more than 1 litre or less than one litre.

3 **a** Give an example of something that is very hot and something that is very cold.

b What is a thermometer?

4 Write the time shown on each clock.

a

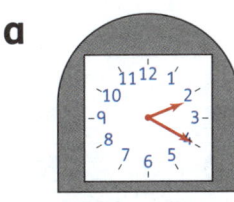

b

c

d

5 A spinner always begins with the arrow pointing up.
Which way will it point after:

  a a half turn clockwise

  b a three-quarter turn anti-clockwise

  c a full turn clockwise?

Use the prices in the pictures to help you work out the answers to questions 6 and 7.

6 How much does it cost to buy:

$30 $20

  a the cap   **b** the ball

  c the T-shirt and the robot

  d the bag and the ball

$25 $55

  e the headphones and the T-shirt

  f the robot and the ball?

7 How much change will I get?

$45 $70

  a I pay for the cap with 50.

  b I pay for the robot with 100.

  c I pay for the bag and the ball with 50 50.

Glossary

< sign – 'less than' sign

> sign – 'greater than' sign

× sign (times) – multiplication sign (see *multiply*)

÷ sign (divide) – division sign (see *divide*)

2D shape – a closed shape with length and width; squares, triangles and circles are all 2D shapes

3D shape – a closed shape with length, width and height

A

anti-clockwise – the opposite direction to clockwise

array – objects or pictures arranged in equal rows and columns; arrays can help us multiply

B

backwards – a direction; the opposite of forwards

balance scale – an instrument that measures mass with two pans; when objects in each pan have equal mass, the pans balance

block diagram – a chart where each block represents one item

C

calendar – a chart showing the months and days of the year

capacity – the amount of liquid a container can hold, measured in millilitres (ml) and litres (ℓ)

Carroll diagram – a table with columns and rows, for sorting items that belong or don't belong in two categores

category (categories) – a feature that we use to sort objects in different ways; for example, types of animal – dog is a category

centimetre (cm) – a unit of length used for short measurements; your thumb is about 1 cm wide; there are 100 cm in 1 metre

centre – the middle point of a shape or object

change – the money you get back when you pay for something with more money than it costs

circle – a round 2D shape

clockwise – the direction the hands move around a clock face; the opposite direction is called anti-clockwise

column – objects or numbers in a straight line from top to bottom

compose – to combine parts to make a number; 3 tens and 5 ones combine to make 35; the opposite of decompose

cone – a 3D shape with a pointed end, a curved surface and one circular face

core – in a pattern, the core is the set of terms in the pattern that repeat

cube – a 3D shape with six square faces

cuboid – a 3D shape with six rectangular faces

currency – the system of money used in a country; in the USA people use dollars and cents; in Britain people use pounds and pence

cylinder – a 3D shape with two circular faces and one curved surface

D

data – information that is collected about a topic

date – the day, month and year; 1 March 2000 is a date

day – one of the days of the week, for example, Monday or Thursday; each day lasts 24 hours; there are 7 days in 1 week

decompose – to break a number into parts; we can decompose 35 into 3 tens and 5 ones; the opposite of compose

degrees Celsius (°C) – units used to measure temperature

difference – we find the difference between two amounts by subtracting; the difference between 4 and 6 is 2

digital clock – a clock that shows the time using numbers and dots; the digital time 07:00 is the same as 7 o'clock

digit – one of the numbers 0 to 9

direction – the way something is moving or facing; left, right, up, down, clockwise and anti-clockwise are all directions

divide – to share or group a number into equal parts; 15 ÷ 3 (fifteen divided by three) is the same as 15 shared into 3 equal groups

down – a direction; the opposite of up

E

edge – a line on a 3D shape where two faces meet

equal – having the same value or quantity

equivalent fractions – fractions with the same value, such as $\frac{1}{2}$ and $\frac{2}{4}$

estimate – a sensible guess; an amount calculated using rounded numbers

experiment – a test to find out the outcome (what will happen)

F

face – a flat 2D surface of a 3D shape; for example, a cube has six square faces

fact family – a set of related facts that use the same three numbers; here are two fact families:
2 + 3 = 5, 3 + 2 = 5, 5 − 3 = 2, 5 − 2 = 3
2 × 3 = 6, 3 × 2 = 6, 6 ÷ 3 = 2, 6 ÷ 2 = 3

forwards – a direction; opposite of backwards

fraction – a part of a whole; one-half, one-quarter and one-third are all fractions

full turn – turning all the way around to return to the starting position

G

gram (g) – a unit of mass used for light objects; a gram is about the mass of two paperclips

growing pattern – a pattern where something is added every time the pattern repeats; the pattern grows from term to term

H

half (halves) – when we share a whole equally into two parts, each part is one-half ($\frac{1}{2}$)

half ($\frac{1}{2}$) turn – turning halfway around to face the opposite direction to the starting position

hexagon – a 2D shape with six straight sides

hour – one hour is 60 minutes

I

impossible – an impossible outcome is something that cannot happen

K

key – a key tells us what each picture represents in a pictogram

kilogram (kg) – a unit of mass used for heavy objects

L

leap year – a year when February has 29 days instead of 28 days; there is a leap year every 4 years

left – a direction; opposite of right

length – how long something is, or the distance between two points

line of symmetry (also *mirror line*) – when we fold a symmetrical shape along this line, both sides fit onto each other exactly

line symmetry – a shape has line symmetry if we can draw a line through the shape to divide it into two parts that match exactly

litre (ℓ) – a unit of capacity used to measure how much a large container holds; a carton of milk usually holds 1 litre

M

mass – how heavy something is, measured in grams (g) and kilograms (kg)

metre (m) – a unit of length used for longer measurements; there are 100 cm in 1 metre

metre stick – a ruler that is 1 metre long, used for measuring length

millilitre (ml) – a unit of capacity used to measure how much a small container holds; there are about 20 drops in 1 ml and 5 ml is 1 teaspoon

minute – there are 60 seconds in one minute; 60 minutes make 1 hour

mirror line – (see *line of symmetry*)

month – about 4 weeks or about 30 days; a year has 12 months; January is a month of the year

multiply (also *times*) – to repeatedly add a number; three times ten, ten multiplied by three and 3 × 10 are all the same as 10 + 10 + 10

N

non-standard unit – an everyday item that we use to measure something; for example, paperclips are non-standard units, centimetres are standard units

number line – a line that shows numbers in order

O

octagon – a 2D shape with eight straight sides

ones – the digit in a number that represents ones; for example, in 345 there are 5 ones

order – to arrange things by size, amount or value; we can order numbers from smallest to greatest or greatest to smallest

ordinal number – a number that tells us the position of something; for example, first, second, third (1st, 2nd, 3rd)

outcome – something that could happen; one result out of all possible results

P

pattern – a set of objects, shapes, pictures or numbers that are organised following a rule

pentagon – a 2D shape with five straight sides

pictogram – a chart that uses pictures to show information

place value – the value of each digit in a number is shown by its place; for example, in 64 the '6' has a place value of '6 tens'

place-value table – a table that has a column for each place value in a number; for example, a tens column and a ones column

polygon – a 2D shape with straight sides

possible – if an event is possible, it could happen

product – the answer when we multiply one number by another; the product of 2 × 2 is 4

property (properties) – a fact about a shape or number; here are two examples of properties: 'a triangle has three sides' and '2 is an even number'

pyramid – a 3D shape with a flat base and three or more triangular faces that meet at a point

Q

quarter – one of four equal parts of a whole; four quarters together make one whole; we write $\frac{1}{4}$ or one-quarter

quarter past – 15 minutes after the hour; 7:15 is quarter past 7

quarter to – 15 minutes before the hour; 6:45 is quarter to 7

quarter ($\frac{1}{4}$) turn – a turn that makes a square corner; four quarter turns in the same direction make a full turn

R

random – not following a pattern, order or rule

rectangle – a 2D shape with four straight sides and four square corners

reflection – a mirror image

regular polygon – a 2D shape that has straight sides that are all the same length

repeating pattern – a set of objects, shapes, pictures or numbers arranged so they repeat again and again

result – the answer to a problem or outcome of an experiment

right – a direction; opposite of left

round – to change a number to make it easier to work with; for example, 18 rounded to the nearest ten is 20

row – objects or numbers in a straight line from left to right

ruler – an instrument for measuring lengths

S

second – when a clock ticks, each tick is one second; there are 60 seconds in 1 minute

side – one of the lines of a 2D shape

sphere – a round 3D shape; a ball is a sphere

square – a 2D shape with four square corners and four straight sides that are all the same length

standard units – official units of measure that everyone uses; for example, centimetres and metres

symmetrical – a shape or object that has a line of symmetry

T

tally mark – a small mark used to count one object; we draw tally marks in groups of five, like this: ⊮⊬

tally table – a table we use to record data using tally marks, to make it easier to count items

temperature – the measure of how hot or cold something is

tens – the digit in a number that represents tens; for example, in 345 there are 4 tens

term – each number, shape or object in a pattern is called a term

thermometer – an instrument for measuring temperature

third ($\frac{1}{3}$) – if you share a whole into three equal parts, each part is one-third

third (3rd) – an ordinal number; the number 3 in position order; we say, 'first, second, third, fourth'

three-quarter ($\frac{3}{4}$) turn – three quarter turns in the same direction make a three-quarter turn

times – (see *multiply*)

times table – a set of multiplication facts for a number; for example, the 2 times table

triangle – a 2D shape with three straight sides

U

up – a direction; opposite of down

V

value – how much a number or item is worth; in 15 the digit '1' has a value of 10

Venn diagram – uses circles to sort objects, shapes or numbers into sets; where the circles overlap shows things that are in both sets

vertex (vertices) – a point where the sides of a 2D shape meet, or a point where the edges of a 3D shape meet; for example, a cone has one vertex, a cube has eight vertices; vertices are also called corners

W

week – there are 7 days in one week

weight – we use weights on a balance scale to measure the mass of other objects; weight can also mean mass

whole – all of something; all the parts

width – the distance from one side to the other; how wide something is

Y

year – 365 days or 12 months